ABOVE REPROACH

A GUIDE FOR WOLF HYBRID OWNERS

ABOVE REPROACH

A GUIDE FOR WOLF HYBRID OWNERS

First Edition
RUDELHAUS ENTERPRISES
Gallup, New Mexico
1995

First Edition, 1995

Published by Rudelhaus Enterprises
Post Office Box 1423
Gallup, New Mexico 87305
(505) 863-5408

ISBN 0-9623640-1-0
Library of Congress Card Catalog Number: 95-70444

Printed in the United States of America

Front Cover: *"Kody" at one year of age. Reported 98% British Columbian/Eastern Timberwolf/Malamute: Owner/Photographer: Cliff and Judy Hoydic*

Back Cover: *Innovative and beautiful security lighting of Wolf Hybrid compounds. Photographer: Lowell and Linda McGinnis*

TABLE OF CONTENTS AND CONTRIBUTING AUTHORS

*Where another author is not specifically listed, the material was authored by Dorothy Prendergast.

PREFACE

Above Reproach is designed to help owners, both first time and experienced, to be the most responsible and knowledgeable owners they can be. It is not intended to, nor will it duplicate information previously published in the book *The Wolf Hybrid,* and readers are encouraged to obtain that publication for even more specific and supplementary information regarding Wolf Hybrids.

Drawing from articles and topics most requested by subscribers to *The Wolf Hybrid Times* magazine, this compilation explores behavior, training techniques, responsibility, legal considerations and laws in effect as of June 1, 1995, containment solutions and other related topics of which all wolf and Wolf Hybrid owners should be knowledgeable.

Drawing from their fifteen years plus of experience and studies, the authors share their collective knowledge in a manner designed to promote responsibility and understanding in a world that sometimes seems to have gone mad.

PROLOGUE

Each year, thousands of Wolf Hybrids are confiscated and euthanized simply because *they are* Wolf Hybrids or suspected of being one. Throughout the years, *The Wolf Hybrid Times* has printed many recountings by the owners of the sad circumstances and reasoning surrounding such confiscations.

The reasoning revolves mainly around the fact that no scientific testing according to USDA protocol has ever been conducted to prove whether the existing canine rabies vaccines are or are not effective in preventing the acquisition of rabies by vaccinated wolves or Wolf Hybrids. The result is that if a Wolf Hybrid bites or even scratches someone, whether vaccinated or not, it will immediately be euthanized and its brain sent off to the laboratory for testing. Owners will be charged for the expense of such testing, as well as held liable for possible prosecution and civil lawsuits.

We begin this book with reprints of some of the articles referred to above in the hope of reinforcing the thought that owners of wolves and Wolf Hybrids need to be more caring, conscientious, knowledgeable and responsible than owners of any other type of canine. This includes exercising greater attention to containment and prevention of access by others to your Hybrid; training, knowledge about behavior, medical information, nutrition, travel considerations, knowledge of legalities of ownership, and more.

ABOUT THE EDITOR

Owner and breeder of Wolf Hybrids for more than twenty years, the author and her family share their northwestern New Mexico home with a pack of Wolf Hybrids that has ranged in number from four to as many as twenty-one.

Prendergast has authored numerous articles which have appeared in periodicals and newspapers nationwide, and, since 1986, has been the editor/publisher of the magazine *The Wolf Hybrid Times,* a magazine about wolves and Wolf Hybrids distributed internationally by Rudelhaus Enterprises.

Prendergast also authored the first edition and revised edition of *Th Wolf Hybrid,* a book that has often been described by others as "essentia for any Wolf Hybrid owner or potential owner."

Rudelhaus, as one would expect, loosely translated from Germa means a "house or place inhabited by a pack of wolves," in this case, Wol Hybrids.

"Moon Dancer." Photo: Jeanne McCarthy

CHAPTER I. TRAGEDIES

DEATH BY VIRTUE OF BEING A WOLF HYBRID

Randal Bowen

A few months ago, I was propositioned to buy a male and a female Hybrid and their litter of cubs. After debating on the financial aspects, I decided to give it a try.

The male, named Titian, was near pure wolf and his behaviors were as a pure wolf. He was a beautiful silver, weighing about 130 lbs. The owner suggested sedating Titian to move him and I agreed that this was a good idea. We went early on a Saturday morning to pick them up. When I arrived, we were told by the owner that shortly before our arrival, a friend of his had come over with his daughter. The daughter had been petting Titian and had been bitten.

Titian was not vicious and was generally shy toward strangers. But he did love to be petted by women and children—so much so that sometimes if one was petting him and pulled back, he would grab at their sleeve and pull their hand back for more petting. He had never bitten anyone or hurt anyone in any way.

But on this occasion, the little girl petting him started to pull away and he grabbed her sleeve to pull her arm back. In the process, he scraped her arm with his tooth. This was a scrape, not a bite. The man took his daughter to the doctor for a shot. This led to a phone call from the hospital to the animal control agency. They came out and picked up Titian. He was partially drugged, but got too excited for the drugs to take good effect.

It was not a pretty sight seeing him treated the way he was. Titian had not had his shots and they had to take him to test him for rabies. We were told that he would be quarantined for ten days, but the next day the owner got a call saying that officials believe that a wolf can carry rabies and not show it the way a dog does. Because Titian was wolf, he had to be destroyed and his head was to be sent to Raleigh to be examined. To me, this was simply unreal.

I do believe that animals that viciously attack a person without cause should be disposed of, but Titian did not have rabies and everyone concerned knew it. It is our responsibility as owners to keep our animals vaccinated. These animals do have a right to live.

We have been hearing a lot about officials that are trying to outlaw Hybrids because there has not been any study done on rabies in wolves. They are saying wolves or wild animals in general can carry rabies without showing symptoms. They are saying they don't know if the present rabies vaccinations have the same effect on wild animals and they consider Hybrids of any percent to be wild animals.

"Loki." Photo: Jan Volkman

CATCH 22 CAN HAPPEN TO YOU

Wendy Hausen

One day a friend of ours accidentally placed Cody, our nine month old male Wolf Hybrid, in a part of our yard from which he escaped. We were away from our home at the time. Cody roamed nearby Bidwell Park and was spotted by our Chico City Animal Control Officer. She called Cody over and he bounded gleefully over to the officer with tail wagging. The officer proceeded to place a wire noose around his neck and one around his muzzle in order to pull him into the back of the animal control vehicle. Cody began to growl and the officer felt the need for further restraint. She and her partner placed a leather muzzle over his head, which he readily removed and gave the nearby officer a fear bite. The bite was minor, but for safety precautions, required medical attention. No stitches were required.

Later that afternoon, we received a call from animal control informing us that our dog was caged in quarantine because he had bitten. We talked about the incident at length and when I was asked, "What kind of dog is he?" I readily admitted that he was a Wolf Hybrid. What a mistake! Why should I lie? Aren't Wolf Hybrids legal in California? I was told by our breeder that they were if they weren't a first or second generation Wolf Hybrid.

There he ate, defecated and was not to be touched. He remained one month in virtual isolation. We were informed by the city animal control officer that Hybrids are recognized by the Department of Fish and Game as an animal that does not require a permit; however, "it has not yet been proven that a rabies vaccine works successfully on Wolf Hybrids." This does not mean that a rabies vaccine does *not* work on the Wolf Hybrid! Cody had current, up to date rabies shots. We knew he didn't have rabies, but yet they followed the procedures for a rabid animal.

Later, we were informed that Cody was declared a "wild animal" by Butte County because of his wolf heritage. The restrictions placed on Cody were as follows:

1) No longer than a 3 foot leash when off your property.
2) A muzzle on at all times when out of his kennel.
3) A kennel area that is to be inspected and approved by the Department of Health.

4) A wild animal permit is now required.

We continue to agonize over the fate of our "dog." We know Cody will not take to a muzzle and will have to be imprisoned on our quarter acre backyard lot. We face a $260 to $1,500 fine for his running at large and biting, even though the animal control officer admitted later in a phone conversation, that the altercation "was provoked."

My recommendations to the readers are as follows: *Never* admit to having a Wolf Hybrid. Do not tell your best of friends that your dog has a little wolf in his blood. (The word may get around and you may be next!)

Further warning: I have now been informed that any Wolf Hybrid that bites in the state of California will have a *six month quarantine.* Our 30 day quarantine was the last of its kind. Butte County residents and Chico City residents should be especially cautious.

REBEL—A NEW YEAR'S TRAGEDY

John Hensley and family

I would like to tell you a story about my beloved Rebel, a Wolf Hybrid. To start with, I need to explain a little about myself. My name is John. I am 28, not married nor do I have any children. I live at home with my parents in Virginia. My older brother, his wife and his two boys, ages 8 and 4, live with us.

We had a male German Shepherd for twelve years. When he died, it took us five years before we decided to get another dog. I had a friend who had a Wolf Hybrid and he seemed to be a very good dog.

When I saw an ad in the newspaper for Wolf Hybrids for sale, I called. The man told me he still had some for sale. My parents and I went just to look at what the man had for sale and we fell in love with those puppies right away.

Rebel was born February 8, 1991. We were told he was a mix between a German Shepherd mother and an Arctic Tundra wolf father. He was just three weeks old when we brought him home. He became part of our family from the beginning and won everyone's hearts. We treated him as most people do a baby; we blended his food in a processor and warmed his milk in the microwave. To me, he was like a brother, a friend, a constant companion. He slept in my room on the bed.

Of course, he rapidly grew, as most babies do, and soon had to start sleeping on the floor beside me. He shared everything with us, our ups and downs, birthdays, Christmas, etc. We celebrated his birthday and Christmas just like a child, with cake, candles and presents.

Rebel also had a fenced in back yard and we were always very careful not to let him out in front of the house. He had the run of the entire house. In the summer, he spent his time laying on the air conditioner floor vents. By now, he had taken over the sofa in the living room as his bed. In the three years that we had him, Rebel was never left alone. Someone was always at home with him. Even though he had grown very fast, he was always treated as if he was everybody's baby.

As soon as Rebel was old enough for his shots, we took him to the vet's. He had all the shots that were required, including rabies vaccination and some that weren't. She never told us at the time that the rabies vaccination had not been proven to work on Wolf Hybrids. We also took Rebel to a dog obedience class which he passed with flying colors.

December 31, 1993, my whole life changed, along with my family's. Rebel was watching from his yard as my brother and nephews were sled riding down a hill with the neighbor's kids. Rebel wanted to get out and go play with them and he somehow managed to open the latch to his gate with his nose and crossed the front yard to go where the kids were playing. He came up on a neighborhood boy and jumped on his *right* rear side, knocking the boy forward to his hands and knees. He then nipped the boy's stocking hat off his head and ran with it. He wanted the boy to play his favorite game with him—"Ha ha, I've got your hat, come and get it." He ran on up the hill and never went near any of the other children playing. The boy just got up without a word or making a sound and walked up the hill to his house.

As we were still chasing Rebel, the boy's father came down the hill and told us that Rebel had bitten his son. He wanted to know if it was that "wolf dog" that had bitten him. My dad just told him that Rebel was a Shepherd mix. He also told him to take his son to the hospital and we would pay for any expenses.

By the time they returned from the hospital, everything had gone crazy. Animal control had called and said that they would be out that evening. When the animal control warden showed up, everyone was worried and scared. We weren't sure what he was going to do with Rebel. The warden scared. We weren't sure what he was going to do with Rebel. The warden filled out a quarantine paper and allowed us to quarantine Rebel for ten days here at home. He told us the boy only had a single puncture wound in his *left* forearm. The wound did not need any stitches so it was just cleaned and a bandage put on it. He also told us there were no rips, tears or holes in the sleeve of the boy's jacket. He just shook his head when we told him about Rebel jumping on his *right* side and said he didn't understand how it could be possible.

My mother called and talked with the boy's father that night and the only thing he was concerned with was that he feared for the safety of the neighborhood. We didn't understand this comment because as far as we know, this man had never even seen Rebel before. Rebel had not been out of our yard since he was a puppy, and then it was only twice. Each time he got out the front door, we were right there to catch him.

Wednesday, January 5, 1994, the animal control warden called and told me that he needed to come and get Rebel to do some testing on him. I asked how long that would take. He just replied that he would explain when he got here. When he arrived, he showed me and my mom a memo from the County Health Director to the administrator of the animal control office saying that the Wolf Hybrid was to be picked up and euthanized. He told us that the reason that he didn't take Rebel from us on Friday was that no one at the animal control office knew the policy that if a Wolf Hybrid bites someone, it has to be taken and euthanized.

We, of course, refused to let them take Rebel. I just couldn't let them have my baby without putting up a fight for him. Since the County Health Director was out of town until the next day, the warden said that he would put if off until the director came back.

My mom immediately got on the phone trying to find out what we could do to stop them from taking my baby. She called a breeder we knew, who got us in contact with the Wildlife Education and Research Foundation. With their help, we branched out from there and were in contact with Dr. Ken Podkonjak, Dottie Prendergast, Carol McKinney, etc., and mom also got a lawyer. Through telephone conversations, faxes, mailing and messenger service, everyone The bombardment of information was outstanding, overwhelming and very informative, but nothing seemed to work. We were on the phone till late into the night and back on it very early the next morning.

The animal control warden and the Health Director both told us that if Rebel had been any other breed, then he would only have been quarantined. But since he was suspected to be a Wolf Hybrid, there was nothing we could do. The Health Director also said if the boy's parents would consent to having rabies shots for the boy, they would not have to take Rebel. Of course, they refused.

Thursday, January 6, 1994 at about 5:00 p.m., two county police officers and two animal control wardens pulled up in the driveway with a search warrant to take Rebel. This was a very emotional time for all of us. My brother couldn't stand the thought of them putting a noose around Rebel's neck to take him to the truck and since I couldn't do anything but cry, my brother took Rebel out to the truck and put him in. Rebel never growled, snapped or anything like that at the strangers that came to take him away. He was actually rather friendly towards them. He probably thought he was just going for a ride.

I practically had to be pulled away from the truck. I didn't want to let my "baby" go. It was horrible. I kept crying and trying to make sure Rebel knew that I loved him. I kept saying over and over to Rebel, "I love you baby," as my dad walked me away from the truck. Rebel possessed such a human personality and human characteristics. He seemed to understand what was going on. He had been pacing the floor and looking out the window every time a car went up the road all day.

After they drove away with Rebel, we got on the phone again in a last attempt to save Rebel. We phoned the County Health Director. He said he wouldn't do anything before 10 a.m. the next morning to give us a chance to talk with the lawyer. We were on the phone till well past 11:30 that night.

Friday, January 7, 1994, my mom was on the phone again starting about 7:00 a.m. The County Health Director seemed determined that despite whatever we could come up with, there was nothing that could be done to save Rebel as long as he could consider him a health issue. At about 9:00 a.m., Rebel was destroyed, his head severed from his body and shipped to a lab for testing.

At about 4:30 that evening, we received a call saying if we wanted to pick up Rebel's body we could, since the test had come back and it was *negative*. My brother, mom and I went to pick him up so that he might at least have a decent burial.

When we picked him up, everyone there was trying to avoid us. They brought out Rebel's decapitated body in a trashbag on a stretcher and flopped him into the back of the car like a sack of potatoes. Then one of the other guys who worked there came out with another bag that had Rebel's head in it and said, "Uh, does this go too?" You would have thought that in their occupation, they could have shown a little more compassion. There was no reason that his head could not have been reattached to his body or at least been put in the same bag.

This shouldn't be happening! A member of our family was taken from us and we were unable to help him. And we still, as of yet, haven't seen the medical report that determined the puncture would have to had been caused by a dog.

I just wish this was all a bad nightmare and I could wake up and my beloved Rebel would once more be with us.

"Tashina." Photo: Shirley Anderson

DOG OR WOLF HYBRID
One Life Taken, Another Spared Through MIs-Identification

Ron White

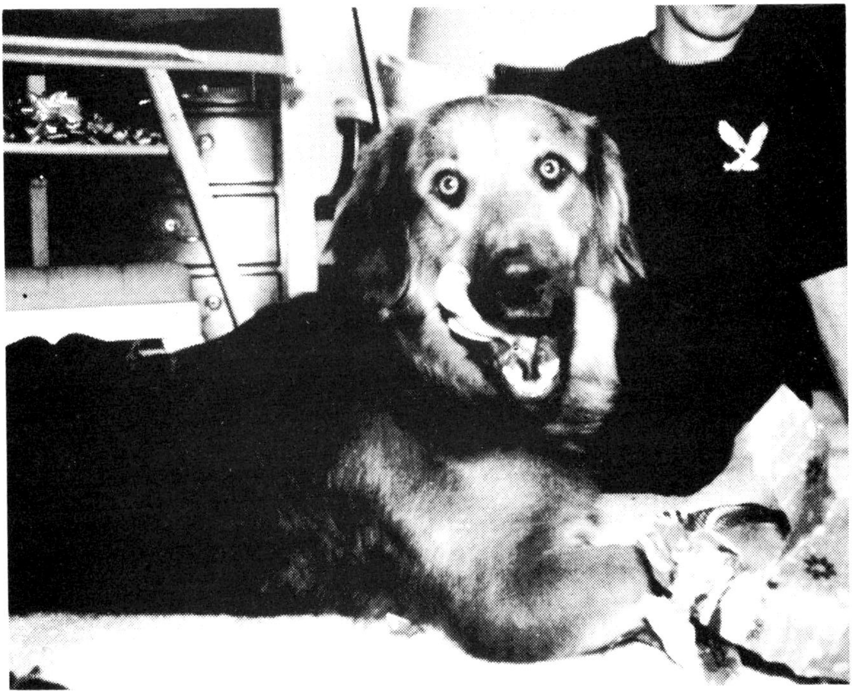

"Tucker."

"Last Christmas, Tucker was a happy camper, opening his presents under the tree with James Moreland and the rest of the family, including the young man's four younger brothers and sisters. Today the dog, described by Moreland as a Shepherd/Husky/Blue Tick Hound mix, is a statistic slated for the ranks of Wolf Hybrids euthanized because the rabies vaccine has not been approved for Hybrid use. But Tucker's family stands by their claim that their pet was not a hybrid.

"We want people to know this can happen to you," said Moreland. Tucker's rabies test came back negative, "like I knew it would."

Moreland said he had to go to three places before finally getting hold of the test documents.

Look closely at the photo reprinted above. Does Tucker look like a Wolf Hybrid to you? This luxuriously coated brown and black dog with its long, soft, hanging ears died in Manassas, Virginia on March 30, 1994 for suspicion of being a wolf/dog mix. The only basis for such a determination were some braggadocio statements made previously by owner James Moreland to friends that Tucker might be part wolf.

But the history of those statements, as published in an article by Vonda K. Wolcott in the *Journal Messenger* was that "his girlfriend told him she had learned from (Tucker's previous owner's) ex-wife that Tucker was a "hybrid." Moreland admitted on the witness stand that he thought it was "cool to tell others that Tucker was part wolf."

Another witness, Dr. Elaine Lutz, Tucker's veterinarian, testified that her records show that on two out of about 16 visits, Tucker's breed was categorized as part wolf. She also testified that "it was fashionable a few years ago for dog owners to claim their pets were part wolf, even if no proof of such heredity existed."

The case began when Tucker bit a ten year old boy less than two weeks before. The bite was not an unprovoked attack. The boy was bitten when he tried to pet the dog after being warned not to. The dog was leashed in Moreland's back yard and had growled at the boy when he approached. Tucker was quarantined for ten days of observation following the bite, and still appeared to be healthy at the date of his death.

But Health Department officials sought to euthanize and test the dog after obtaining copies of his veterinary records. They said "domestic dogs are spared from being tested for rabies because history shows if they remain healthy ten days after biting a person, then the person is assured he is not infected. However, hybrids such as dogs crossed with wolves or coyotes do not necessarily show signs within this quarantine time and must be euthanized and tested, according to state law. This testing must be done even if the animal has been vaccinated against rabies because the vaccine used on domestic dogs is not assured of preventing rabies in hybrids," the official said.

On March 29, 1994, Judge LeRoy Millette Jr. ordered that Tucker be euthanized and his brain tissue tested before the ten year old victim began the third in a series of prophylactic shots. Tucker was euthanized on

March 30th, the morning Moreland had made plans to visit one last time and say goodbye to his beloved pet.

But it wasn't to be. When he arrived at the shelter, unbeknownst to Moreland, Tucker had already been put to death and his head taken for examination. On the afternoon of the 30th, the report came back negative. Moreland said he was totally unfamiliar with the term "Wolf Hybrid." "He was a beautiful dog. I had no idea that what I thought was a compliment to my dog would stamp the animal as a Wolf Hybrid and make me feel responsible for its death."

Tucker was buried March 31st in a purple box with all his toys and his leash. "It rained the day we got Tucker and it rained the day we buried him," said Moreland's mother. The entire family and a friend were there to say goodbye as Tucker was laid to rest on top of Bull Run Mountain with a beautiful view of the Blue Ridge, one of his favorite running spots.

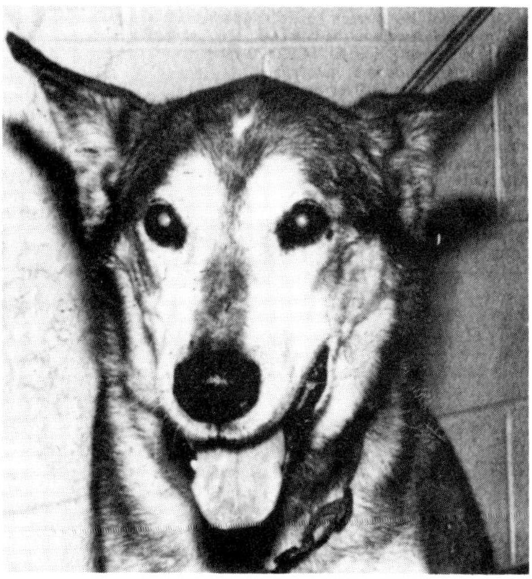

Meanwhile, the animal shown above was picked up by the authorities as a Wolf Hybrid. The owner claimed that it was a Shepard/Husky/Retriever mix and the animal was released to him. Compare this photograph with that of Tucker. If these two animals were placed side by side and you were told that one of them was a Wolf Hybrid, which one would you pick as being "part wolf?" This animal or Tucker? Compare the phtographs following.

Wolf hybrid? No! It's a huskie from the shelter who was identified as having an owner who would redeem him.

A German shepard waits at the county shelter for someone to adopt him. A pleasant animal, but scared of his present surroundings, he was given up by his owner because he was not housebroken and "eats everything."

There is a strong message here, and that is that *no dog* is safe from confiscation if even one misinformed person *thinks* it might be part wolf or part coyote. The animal does not even have to be involved in a biting incident

If it ends up at the animal shelter, in most areas its sentence is automatic. And in all areas, if the animal has bitten someone, its death can be expected before any legal action can be instituted.

In a very good article entitled "Wolf Hybrids Still Waiting For A Vaccine" which subsequently appeared in the *Journal Messenger* on April 25th, Nancy White presented a well-balanced discussion of the issues now surfacing in the state of Virginia. She presented quotations from animal control officials outlining the difficulties in identification, such as that of the Prince William County Administrator. "Supposedly hybrids have a black spot three inches below the base of the tail. I have not proved it out positive yet. In reality, I do not believe there is a categorical way of identifying them externally."

White goes on to say that "Without a bona fide test, veterinarians and government authorities have to accept what they are told an animal is. It is believed one test is coming within a few years that will identify dogs, wolves and hybrids using blood or other tissue. Such a test would certainly save the dogs, but wouldn't do anything for the hybrids if the dog vaccine does not gain approval for use in hybrids. Until studies prove the efficacy of the vaccine in these animals, the realization of a rabies vaccine licensed for hybrids remains out of reach."

Said Dr. Jared Florence of the Prince William Health Department, "We don't trust rabies in dogs when there is a break in the skin. If the dog is alive and well after ten days, then it's okay. But it's still not known if the ten day (viral) shedding period holds true for wolves and wolf mixes...Nobody is willing to take a chance with a human life."

What should Tucker's story teach the owners of Wolf Hybrids? And the owners of German Shepherds, Malamutes, Samoyeds, Huskies and mixed breeds as well? The message should be loud and clear that so long as the rabies vaccines have not been tested specifically on wolves and Wolf Hybrids, the possible danger to public health places ALL these animals in jeopardy. The key words are *public health,* not breed discrimination.

How many more innocent Tuckers will have to die before enough money can be accumulated by to accomplish this important testing? When can we all live without fear?

CREE

Margo Soulé

Today marks the fourth month that he has been gone. He was killed by a direct order from the SPCA. I still can't believe it and I am continually reliving the nightmare.

I named him Cree. My daughter and I had visited him quite often after he was born and now he was ready to go home with us. What a cute little guy he was. So scared and yet so determined to be brave. He sat in my lap the whole way home. Once inside the house, he backed himself into a corner and growled at anyone approaching him. We left him alone to let him calm down a bit. After all, he was away from his family.

We waited a couple of hours and then approached him. It was a different story. We were welcomed with such enthusiasm—at least as much as a scared little cub could muster. He completely won our hearts right then (as if he hadn't already) and we became his slaves. For the next three nights, I slept right near him on a floor mat. Cree was a never ending source of amusement and I took plenty of pictures to prove it. I teach and was off for the summer, so we spent most of my waking hours together—sleeping hours too, for he made his "den" under my bed. We had thoroughly bonded.

In late July, my daughter and I had planned a vacation. I shopped around to find Cree a suitable place to stay. I wanted one with a run and shelter also from the almost daily summer showers. I dropped him off, little realizing the nightmare to follow.

We returned the first week in August, got off the plane, into our car and headed straight to the vet to pick Cree up. I couldn't wait to see him. We were wondering how much he'd grown and changed in his colors. His tail had gone from a small black tip to being almost totally black before we left.

The receptionist said he was wet and couldn't come out just now. I asked her why and was told he'd been bathed. I was puzzled, for the original order to bathe him had been cancelled by me before I left. She looked at her records and said, "Oh well, I guess you get a free bath." I told her that I wanted him wet. I didn't care—I just wanted to take him home.

A woman vet then came out and said Cree had nipped someone when they had tried to bathe him. I said I was sorry but wasn't too surprised, considering he was just a four month old puppy They still had not brought him out to me.

Then a male vet appeared, saying they would have to keep him there for observation for ten days. I was thinking, "Oh no, here we are gone for six days and now ten more before we can take him. How would that affect him?"

I asked if we could see him. He was in a small drying cage. Boy, was he glad to see us. He was wagging all over. My eyes filled with tears and I was fighting back sobs as I had to leave him there.

The next thing I remember was the vet on the phone with the head of the SPCA. They were discussing having him destroyed because he was a Hybrid. This can not be happening!! I immediately called my breeder who agreed to meet me at the SPCA to try to find out what was going on and to try to prevent the worst from happening.

The breeder arrived first and as I saw him exit, he sadly said, "I think you have lost him." Now it is my turn to go in and appear calm when my insides are crying. I told him that I had been nipped plenty of times and I was fine. He asked me if I had reported it and when I said no, I was told that it was a misdemeanor, but I would not be charged this time. He said I could spend six months in jail or be fined $500.00. There was no point in talking to him anymore. His mind was made up before I had even arrived. As I left, tears were streaming down my face.

After returning home, I began to phone anyone I thought could help me. I called my councilman, I called an Animal Control Officer friend in Florida who was a personal friend. He made some calls and got me the name of a lawyer in Virginia who worked with cases similar to mine. I called another local vet and two other lawyers here. All agreed that it was very stupid and needless. This was all done on Thursday afternoon. By Friday afternoon, they had killed him, cut off his head and shipped it away to be studied. The lawyer in Virginia returned my call that afternoon, but he was too late. It happened so fast.

I am home surronded by his feeding bowls, his toys, and where is Cree? He didn't even stand a chance. They didn's even take into consideration that he was a four month old puppy and had never been out of my yard or my house before. My heart still grieves.

"Tundra." Photo: Debbi McDonough

CHAPTER II. RESPONSIBILITY

This volume may not set so well with some owners. Its primary focus is on knowledge and responsibility—responsibility of all owners and/or breeders of Wolf Hybrids, not just "the other guy."

Responsibility in this context takes many forms. In terms of protecting the reputation and safety of the "breed," it begins with the breeder, whether it be a professional breeder or a one-time breeder. A true professional has, through many successive generations of breeding and culling, reached a point where he or she can with reasonable certainty produce puppies with good temperament but wolfy appearance; will provide a multitude of printed and verbal information (both pro and con) to prospective buyers; will provide detailed information on the animals' lineage, percentage and number of generations removed from a pure wolf and require that a contract be signed by the purchasers of their cubs.

A professional will screen and educate his or her buyers carefully, to insure as much as possible that the animals will not end up being chained, mistreated or neglected, either physically or emotionally, nor allowed to roam free. A professional will not sell the "rejects" or pups that should be culled to anyone inexperienced or unprepared for what they are buying and those culled will be required to be spayed or neutered. All breeders should do the same.

Responsibility applies to the buyer also—responsibility to learn all you can about the animal you have acquired through reading and absorbing every bit of factual information you can find about them, responsibility to provide the best environment for the rest of the animal's life without allowing it to roam at will and suffer the consequences of an early death due to an ill-fated encounter with modern technology (being hit by a vehicle or being shot) or of an unanticipated bite, attack or pillaging incident. Responsibility means never even considering turning the animal out to fend for itself "in its natural environment" where it will most surely die, for the wilds are NOT a Wolf Hybrid's nor a captive bred wolf's natural environment and it is most likely to die of starvation or be killed.

Responsibility means not allowing even an outside chance of a small child (or any other living being) approaching or eliciting prey-type responses from either your Wolf Hybrid or any of your other (perhaps normally excitable) animals. Responsibility means using the same sort of common sense in protecting the animals as one would use in protecting one's own children from the hazards of every day life.

Responsibility means taking an active part in the decisions that are being made "on your behalf" regarding your Wolf Hybrids—some being made by the Wolf Hybrid organizations themselves—some by controlling state and local governments without opposition from most of those individuals or organizations affected until they find themselves in conflict with rules or legislation they did not actively oppose, or propose reasonable alternatives to.

Responsibility means not waiting until your own animal is threatened with confiscation, death or decapitation before helping in any way you can to provide proven, legally recognized protection for all wolves and Wolf Hybrids in all states against rabies (and other diseases).

Responsibility is not an easy pill to swallow (it usually applies to someone else it seems). It is a pill all of us must swallow—and now!

As proud owners of one of the most magnificent and intriguing breeds of canines to be found, we can not allow them to be in jeopardy because "we are certain that our animals would never be the object of bad press, be confiscated or actually bite someone!"

We cannot allow them to be in jeopardy because of our indifference, lack of knowledge, lack of adequate containment facilities or because of personal jealousies or problems we may have with other Wolf Hybrid owners or organizations. We can not afford not to take steps to promote reasonable and uniform containment regulations (of ALL large canines) in order to prevent the Wolf Hybrid from being singled out as a bad breed, as the press is often want to do.

We feel that those who are unaware of or choose to ignore the rash of very serious attacks on children by Wolf Hybrids need to be educated.

"Moon Dancer." Photo: Jean McCarthy

These attacks (recounted in the chapter on containment) could all have been prevented by the use of a little common sense exercised by the owners or caretakers of these animals. Granted, some common sense exercised by the parents of the children who have been injured by canines would have helped, but children are children, and just as parents lock away poisons and medicines to prevent "curiosity from killing the child," so must the children be kept away from animals who in an instant, could kill or maim them. To a one, these incidents might never have happened had some common sense been exercised.

But— there are children who are dead.

But— a few rather elementary precautions and understanding of canine behavior could have prevented it.

But— there is still no proven and licensed rabies vaccine available for wolves and Wolf Hybrids.

But— the animal always loses.

The Wolf Hybrid is in great danger of becoming an outlawed animal on two issues:

1) that of being a potentially dangerous animal and the lack of sérious attention given to this fact by owners; and 2) that the rabies vaccines have yet to be proven effective for wolves and Wolf Hybrids. Both problems can only be resolved by you, the owners and breeders of these animals you love and respect so dearly.

I looked out my back window the other night and saw a young cub on the other side of the fence in a yard that adjoins my property. It was obviously a Wolf Hybrid and not one of the feral dogs that run around the area. It was just as obviously from my own line—one that had been sired by my male some time back.

The owner of the dam lives some miles from me, but the lady to whom this pup and a black female pup had been given, lives less than 1/2 mile from me. I had given this litter their shots from the time they were old enough until their first "permanent shots," and had helped nurse them through all kinds of problems. I felt responsible and obligated somehow. Now, here was one of them—turned out, so to speak. They might as well have been. Though the lady had a 6' chain link fence around her place, the gate was rarely closed. It is not my idea of containment. Each time we gave the pups shots, the lady was so worried about the poor puppies. Now it seemed like such a sham. What do you do in such a situation? You aren't the owner—you're not even the person who sold them to her.

I first called the "owner." She said she had company, but she would come and get the pup immediately. She didn't mention the other little black cub that was raised with this puppy. An hour later, having seen no activity on the other side of the fence, I went over to the neighboring property myself, only to be greeted by a rather hostile father who was adamant that I wasn't going to take his little daughter's new friend away unless I could prove I was the owner and pay for the food he'd bought for it—still unwise to the fact that it was a Hybrid.

It took some tall talking to convince him I was only concerned about the animal's well being, all the while watching "Alpine." as she was now being called, and the man's six year old daughter tumbling about the yard, so obviously enjoying themselves and each other. I finally told the man that she was a Hybrid and given such a family and the containment it needed, there was no way I would call the real owner back again. (She never did show up, by the way). This bit of information only served to make the family more attached to her than ever.

Alpine still lives on the adjoining property and accompanies her young friend wherever she goes. There is also another dog to play with and the family has built them a 6' wide, 3' high "hogan" for shelter. The little black cub, more wolfy in temperament than her littermate, has disappeared and it disturbs me greatly. She never did get her rabies shot.

It is sometimes very difficult to determine where one must draw the line between conscience, concern and responsibility on the one hand, and the legalities of ownership, responsibility and the infringement on other people's rights on the other. Where does "rightful concern" become "illegal interference?" The county animal control officers as well as other landowners may well shoot the black pup for destroying property and/or pets. A private individual, however, may well be liable for theft, harassment and other charges in trying to "save" what they perceive as a mistreated or neglected animal.

A related section discusses different people's versions of "adequate." If an animal gets out, the containment is obviously not "adequate." It is our hope that owners with facilities which would obviously not contain a determined animal will upgrade their facilities, will not rely on the heretofore reliability of their animals, but instead will take preventive measures before they are needed and the animals get out and are lost, stolen, shot or put down by either local animal control officers or Fish and Game. You must remember that these animals are your responsibility—morally and legally. If something happens to them, you'll cry. If they do damage, you'll pay, and probably still end up crying if they are destroyed for what they did

There are many aspects of responsible ownership and breeding. All of the subjects covered in this book relate in one way or another to being a responsible owner—verification of percentages, provision of secure and adequate containment facilities, understanding behaviors and their possible modification, obedience training, obtaining a general knowledge of many other topics affecting this diverse breed of canine, and knowledge of legal requirements which may affect you and your animal.

"Wolfie." Photo: Dean Simonson

INTEGRITY AND HONESTY

Over the past few years, radical changes have taken place regarding requirements for proof of lineage in several of the national registries of Wolf Hybrids. While many owners/breeders have not liked the changes, this writer and many other owners/breeders and observers are greatly pleased by the ongoing effort to restructure the registries to reflect the true percentages based on bloodlines rather than strictly visual evaluations or wishful thinking.

The dissatisfaction stems mainly from people who have purchased animals in the past which were falsely represented as high percentage animals, only to find out, when they are trying to, or have been selling pups, that the animals they have are not what they were represented to be. Is this a matter of money or a matter of pride? It is, in either case, not fair to either the animals or the potential owners.

The U.S. Wolf Hybrid Association underwent major changes in its registry in the mid to late 80s and early 90s because many percentages had been previously accepted based upon the owner's word on percentage and lineage. On investigating, however, it was found that some well known breeders had misrepresented the percentages of their animals, both to the registry and to potential buyers. On deeper investigation, it was found that on occasion, the same animal was credited by the breeder with several different percentages. Proud owners and/or new breeders were also guilty of not only innocently accepting these percentages, but exaggerating them still further.

Iowolfers, Inc. is still going through the same painful process. There are, it has been discovered, many animals across the country which are affected, and though the animal's behavior (and perhaps looks) is, by an owner's or breeder's evaluation, "very wolfy," there is now a question as to whether the "very wolfy" behavior may not instead be attributable to personality problems related to an excessive number of generations of inbreeding.

Iowolfers has refused to issue certificates when they are aware that the percentages may be in error and are strict in demanding traceable documentation on every animal submitted for registration. As a result, they have been harshly criticized as becoming a "police organization."

Isn't that what the Wolf Hybrid community needs at this point in time?

It takes a courageous, honest and caring person to admit that what he has bought/sold is not what he thought, that the percentage is lower, that the degree of inbreeding may be much higher than he had ever dreamed, or that, in some cases, he has been sold a much higher percentage than he thought and perhaps would not be able to deal with. The misrepresentation works both ways when a hungry breeder is looking to get rid of the last of the litters.

Pity the pour soul (first time buyer) who thinks he is buying a 50% Hybrid which is being touted as "very doggy" in personality and comes home with a very shy, fearful or high-strung "very-near-wolf." Pity also the other end of the spectrum—the buyer who thinks he has purchased a very high percentage Hybrid which is actually a 25% or lower, and because the animal is so docile and manageable, thinks he can handle a full wolf. Pity also the children who have been or will be exposed to the idea that pure wolves are gentle and trustworthy because those "super high percentage Wolf Hybrids" they have met have all been that way.

The question again comes back to honesty. Should you find out, for example, that your animal actually has NO wolf blood, you owe it to yourself and the purchasers of any puppies you may have sold to tell them that. Mutts are not in jeopardy, either as being risks for rabies (they would be quarantined for a minimal period of time rather than euthanized), nor as a breed. Should you find out that your animal is a higher percentage than what you thought—you'd better tell any buyers that too! Nowadays, you can be sued for selling anything that is not what you represented it to be. You can also be sued if the puppy gets into trouble, even if you were only the breeder. And if you have a conscience and care...

This editorial is not intended to offend, nor to point fingers. It is simply written to lend support and encourage every effort at honesty among wolf and Wolf Hybrid owners and breeders, to be proud of and love the animals you have owned for years, whether they are what they were represented to be or not, and to be honest to those who meet your animal, look at your puppies, and to those with whom you may register them.

The future of the wolf and the Wolf Hybrid is still very tenuous. Their existence may very well depend on the integrity and responsibility of private wolf and Wolf Hybrid owners. If you, as owners, breeders, admirers of this magnificent animal, cannot be honest, how can you expect that the animals from whom they sprung will be allowed to live?

PEDIGREES

Warren Spangler

As a breeder, I have seen many copies of "pedigrees" issued by other breeders around the country. It disturbs me very much that these so-called "pedigrees" can in no way be verified by a prospective purchaser. I recently received copies of some pedigrees listing some of my own animals with radically WRONG percentages; and I have seen some that would drive a mathematician crazy! Many of those I have seen provide no documentation of the animals' names, let alone names of the breeders or owners.

Although I send out corrective letters regarding my own animals, many of the people requesting such information subsequently refuse to correct their own "pedigrees," and so the subsequent purchasers of their litters write again asking for verification.

I recently received a copy of a pedigree from a new owner in Colorado which would boggle your mind. It runs through five generations, but is almost indecipherable and does not even list a final percentage for the wolf/dog pup or its parents. Besides being almost unintelligible as to what animals were the parents of which animal, if one goes through the entire thing with a calculator, using the figures on the paper, even the earliest generation showing percentages reduces a rather high percentage animal to less than a 25% animal. And, of course, no other information was provided—not even a hint as to how one could begin to trace the true lineage.

What Is Wrong With These Pedigrees?

In the first case, not only were the percentages misrepresented, but even the names of the ancestors were changed. In other words, it is altogether false.

In the second case, the percentages could never have been derived from the percentages of the animals listed in the previous generations—not only in the last generation, but from the very beginning of the pedigree. Additionally, it is impossible to determine who the sires and dams were of any particular animal.

Neither pedigree provides any information for tracing the animal's lineage for verification purposes.

What Should Appear On A Pedigree?

For each animal listed, not only should its name and percentage appear, but if there is a mixture of dog breeds or wolf subspecies, a specific listing of those breeds and subspecies and a means of verifying that information. The name and last known address and/or phone number of the breeder/owner of each animal should be provided. If any of the animals were ever registered with any organization, its registration number and the address of the organization should also be provided. This does not make the pedigree foolproof, but it does at least give a lot more incentive to the breeder to accurately verify the percentages of the animals they are selling, as a prudent purchaser could take the time and effort to check out the pedigree and the ancestors to see if this was indeed the type of animal they were seeking, and if the lineage was indeed accurate. Additionally, people providing misrepresented histories or lineages could very well be liable for fraud.

There is an additional item that I include on my pedigrees to further ease the task of verification. I provide wherever possible, the date of birth of each animal. This enables the owner/breeder of an animal from generations back to check their records (or even recall from memory) a particular litter of pups, as they would likely have no way of knowing what each pup was named once it was sold.

Personally, I would like to see the Wolf Hybrid organizations/registries require that all this information be provided before registering a litter. It would certainly make their jobs a lot simpler. In this day and age of the computer, an animal's lineage and percentage could be verified by number and not involve the tedious tracking of animals/owners they so often face today.

AKC requires most of this information and more. This is just one man's opinion, but if this information could be required by all registries, we'd all be a lot better off.

(Example of a BAD pedigree)

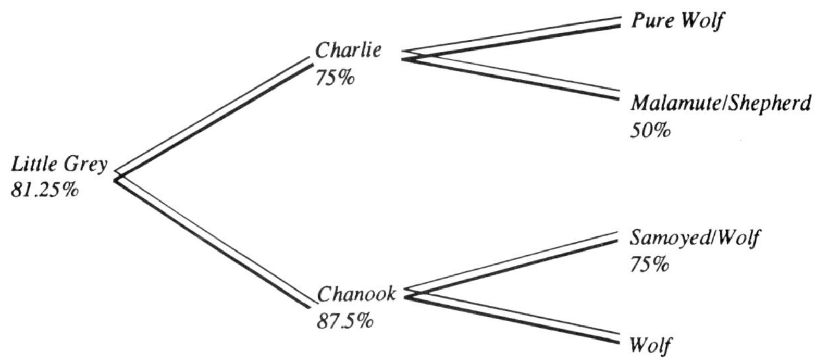

(What it SHOULD look like)

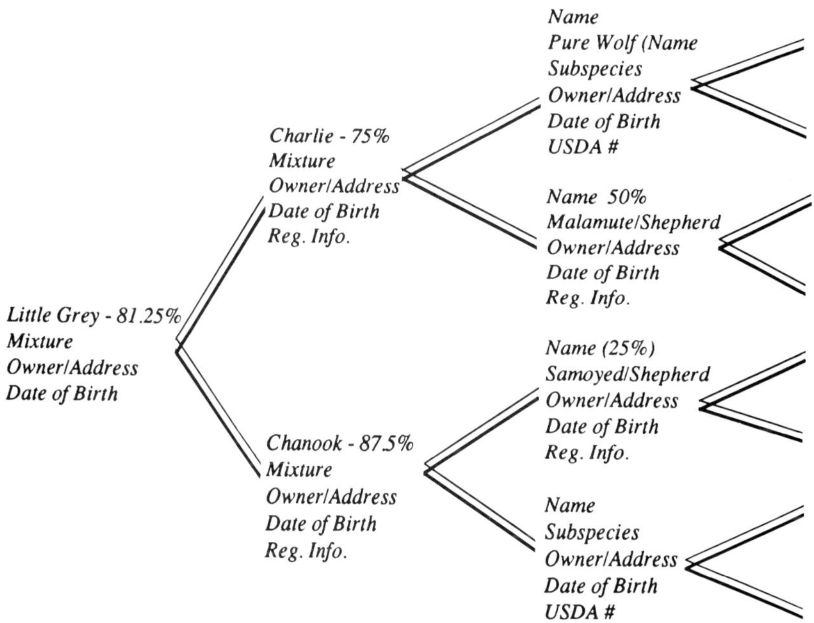

Figure 1

(Simplified examples of bad and preferable pedigrees. Space requirements here limit listing prior generations, but ALL generations which are known should appear on any animal's pedigree).

"Lobo of the Tetons." Photo: Mary McBee

CANINE BREEDING PLANS

Mary McBee

Plan for Excellence

Before beginning, let me state that I am surely no expert on this topic and a short article like this can't help but do injustice to a field of much depth and complexity. Therefore, please consider this a very simple introduction. At the close of this article, I'll mention a few books on the topic that have been favorites of mine through the years, although I'm sure other good publications are also available.

Breeding your dog. It sounds so simple, but is it? What are your reasons? Do you want only one litter to provide pups for friends? Do you want to breed several females and have six litters to sell over the next four to five years? Or do you want to get really serious and devote the next 15-20 years to developing your own line or strain? Answers to these questions make a big difference in what type of breeding plan you choose, if any. But regardless of your reasons, it's something that needs to be given very serious thought.

By choosing two animals to be bred and to create offspring, you are for all practical purposes, "playing God"—you are creating life and you are molding it. As one author said: "If you're going to play God, then do it right or don't do it at all!" Think seriously about what you want to do. Obtain knowledge and then proceed with care, compassion and good purpose - or don't do it at all. This a big order to fill, for us all.

Here are a few simple terms with which we need to be familiar. Definitions vary, so I will use those as defined in Eleanor Frankling's book Practical Dog Breeding and Genetics.

INBREEDING: breeding of very close relatives, such as parent to off spring or brother to sister.

LINE BREEDING: breeding related animals, but those of more distant relationships (i.e., grandfather to granddaughter, cousin to cousin, etc.)

OUT BREEDING: breeding of very distant relatives with no similar ancestors present for at least 3–4 generations back in each animal's background. (i.e., poodle to a poodle when very little known relationship exists between the two).

CROSS-BREEDING: breeding two animals of different breeds, such as a Collie to a Beagle.

OUT-CROSSING: breeding a line bred or inbred to a more distantly related animal with the intent to draw on specific traits on a one-time basis only, and then return to line or inbreeding as a constant once more.

BACK CROSSING: taking a hybrid offspring (usually the result of cross-breeding) back to one of its "pure" parents, i.e. taking a 1/2 Malamute, 1/2 wolf back to its pure Malamute mother.

Most people who want only one litter of pups will look around and pick out a male that looks good to them and will breed him to their female. Hopefully, they will pick an opposite of what they want improved in their female. If she has very short legs, they should look for a longer legged male; if she is smaller and they want more size, they should look for a larger male; if she's a little shy or spooky, they should look for a calm, outgoing male, etc. At the most basic level, one should at least try to avoid finding a mate that has the same "faults" their animal has, so as to hopefully avoid perpetuating or accenting those same faults in the offspring.

A more serious breeder will choose a male for their female by looking beyond the male itself and into its background, thus gaining some knowledge of what gene traits that male is carrying and might throw, regardless of what he looks like. Does he have a pedigree listing names of ancestors for at least three generations back? If so, that's a start—but only a start. Ask for information on those ancestors—the colors and markings of each, photos of each so you can see conformation. Ask about temperament problems, health problems like hip dysplasia, blindness, hearing problems, etc. Take a good look at the parents and grandparents of that male. Some of your pups will probably turn out like them, not the male you've chosen to breed to your female.

In breeding is also practiced on occasion in order to find faults and bring them out into the open so they can be eliminated rather than hidden and passed on. If one breeds a brother to a sister and their litter is composed of two greys, two blacks and one spotted, and if you consider the spotted a marking fault, a conscientious breeder would either cull the spotted pup, or if this is too difficult, neuter it before it reaches sexual maturity, but either way, will make sure this animal does not perpetuate that fault further. The other pups in the litter that resulted from this tight breeding of brother to sister may no longer be carrying this spotted trait, so it might be eliminated from the line immediately (although one is not always that fortunate so quickly). But by breeding intensely, breeders try to bring out faults and eliminate them rather than perpetuate them generation after generation through out-breeding, where it can continue to crop up unpredictably as generations go by.

Please take note here, I've never been one of those who has considered culling a very viable alternative, so I have never been able to consider myself a truly dedicated breeder.

For the most part then, out-breeding is the safest, easiest method and usually produces pretty nice first generations. But offspring from out-breeding are not necessarily real good breeding stock because what they will throw in the future is often not very predictable. Line breeding and inbreeding are more for those serious breeders who want to study the gene pool they have to work with and want to "set type" over a long term basis.

Many of us have a distinct prejudice against breeding close relatives, since we've always been taught that this practice is totally abominable (which in turn happens to preserve social structures in our complex human civilizations). Biologically, there are many pros and cons to the issue. However, pure wolves in the wild have commonly practiced line breeding and inbreeding for thousands of years. Most packs are made up of close relatives and remain so for centuries. Occasionally, one breaks off or is driven out and eventually forms a pack with other unrelated loners, yes—but according to reputable researchers, this is the exception, not the rule. Indeed, wolf packs in the wild have thrived by line breeding and inbreeding, but for one very important reason: Mother Nature has provided the ruthless and severe culling necessary for this type of breeding plan to be highly successful. Many of we humans cannot be so methodical or objective in culling or in selection of breeding stock along the way, so it is a more difficult method for us.

There is also a very common misconception constantly stated in regard to close breeding. How often have you heard someone say, "that breed has been ruined because it has been inbred too much" or "too much inbreeding has ruined the temperament of that breed and caused them to be too high strung?" Any time any breed becomes popular and there's a demand (market) for the pups, poor breeding will result because too many people who either are not serious or who don't realize fully what they are doing (or don't care), are breeding poor quality animals and allowing them to reproduce. These animals fall short in either temperament, conformation, intelligence, or in the genes they are carrying and perpetuating for these traits. The way they are bred is usually not the cause of deterioration in quality of offspring; they often should never have been allowed to reproduce at all, regardless of how or what breeding plan was chosen. Or at the least, if bred, breeding plans should have been selected that would have helped eliminate those poor genes in offspring in the future. For those who would like more information on breeding plans, the following would be helpful:

Practical Dog Breeding and Genetics by Eleanor Frankling, Arco, Pub. Co., New York. (Good general information on breeding, whelping, basic genetics, breeding plans).

The New Art of Breeding Better Dogs by Kyle Onstott, Howell Book House, New York. (Good general information, more complex genetics, preference to line breeding and inbreeding plans).

Planned Breeding by Lloyd Brackett (small booklet), Dog World magazine, Chicago, IL. (Compilation of articles written for Dog World. Mr. Brackett's development of top quality line of German Shepherds. Definite preference to line breeding and inbreeding and explanations why).

"Kweo." Photo: Joan Fleetman

SALES CONTRACTS

Mel Tanner

The Wolf Hybrid Times receives many queries about sales agreements, whether they are needed, if they are binding, and for samples of agreement forms.

Breeders should always require the signing of a sales agreement or contract prior to sale. Such contracts not only protect the seller, but also the purchaser.

Any number of conditions may be put into the contract—conditions which must be agreeable to both parties. In order to be certain that the contract is legally binding, it must be signed by both parties and dated before a Notary Public and bear the Notary's seal. The effect of this is to create a contract which can be recorded at the County Clerk's Office, should such a need ever arise (as in the case of a dispute between the parties).

The form of contract which follows can be modified in any way to suit the parties' needs. Clauses can be deleted or others inserted; however, it is very important that the notarization requirements remain, for the protection of both parties.

You may wish to change the method of payment; however, I do not recommend a time payment plan unless the amount is paid in full by time of delivery of the animal(s). Far too many breeders have come up empty-handed by agreeing to payments over a period of time after the animal has already gone to its new home.

I would definitely recommend the right of inspection of the purchaser's facilities prior to the sale of the animal(s). Don't just take the purchaser's word for it—it may come back to bite you if the animal gets into trouble for lack of adequate facilities, just as it will the purchaser.

Always provide the purchaser documentation of the animal(s) lineage and percentage, certified to the best of the breeder's knowledge—and this must be delivered to the purchaser at the time of delivery of the animal(s). From the purchaser's point of view, this is essential, as too many people have relied on "what the breeder said," or on being promised the documentation later, never to receive it. This can cause many hardships to the purchaser who may not be able to register his/her animal(s), or who may have to prove their percentage to animal regulatory agencies.

Make no mistake—sales agreements or contracts are important to both buyer and seller, and both parties should insist on having one which protects all concerned.

"Leila." Photo: Janet Lux

SALES/PURCHASE CONTRACT

This Contract, dated ______________________________, by and between ______________________ whose address is ______________________________ ____________________ (Seller) and ______________________________________ whose address is ____ (Purchaser) constitutes a valid and binding agreement under the laws of the State of ____.

In consideration of receipt of the full price of ____________________ ($__________) Dollars, Seller agrees to sell and Purchaser agrees to buy the following animal(s):

The foregoing purchase price shall be paid in the following manner: ____ ($ ________________) on the ____________ day of __________________, 19___, and the balance of ________________________________($ ____________) on the date of delivery.

Both parties agree that the purchase price for the animal(s) will be non-refundable after 30 days from the time of sale, whether or not the animal is returned to Seller, for any reason.

The conditions of sale and acceptance are as follows:

1. Prior to delivery of the animal(s) to Purchaser, Purchaser will provide a secure and comfortable yard area of a minimum of 1/8 acre, with no more than two animals kept in this area at any time. The yard area will be strong and secure enough to prevent escape of the full grown animal (farm fencing is inadequate. Heavy gauge chain link is preferred). Fencing must be a minimum of six feet high. If this height is not sufficient to keep the animal contained, either the fencing will be heightened with security wire slanted into the pen (not barbed wire) and/or an electrically charged fence wire will be installed inside the fence approximately one foot from the vertical fence and just at nose level of the animal(s). Secure skirting will be attached to the inside bottom of the vertical fence to prevent digging out. If the

fencing is inadequate, the cause for the inadequacy shall be remedied immediately and all actions of the animal(s) when not in its containment facility shall be the responsibility of Purchaser.

Purchaser also agrees to assume full liability and responsibility for any person or animal which enters his property, whether an invitee or not, for any damage which the animal(s) may cause.

Seller has the right to inspect the facilities prior to delivery of the animal(s).

2. The animals described above shall NOT be restrained on a chain or tie-out, except on an emergency or very temporary basis, not to exceed six (6) hours, and then only in a sheltered area where it will be protected from intruders or vandals and inclement weather.

3. A shaded area and waterproof shelter shall be provided by Purchaser for each animal at all times within any area where the animal(s) is/are contained.

4. Clean, potable water shall be available at all times; nutritious food will be provided in sufficient amounts on a daily basis, and all vaccinations will be kept current and updated. Any and all veterinary care deemed necessary to keep the animal(s) in good health shall be provided by Purchaser.

5. The pen (or other area where the animal is kept) will be cleaned of droppings on a regular basis to prevent disease.

6. The animal(s) will have human companionship for at least two hours per day, or will have a compatible canine companion at all times. Both are preferred.

7. If the animal(s) is/are female, she/they will be spayed by the age of six months, unless this is waived in writing by Seller. A copy of the spay certificate shall be sent to Seller as proof of

compliance. If this condition is waived by Seller, the femaleshall not be bred sooner than two years of age and no more than once per year.

8. If the animal(s) is/are male, it/they must be neutered by the age of nine months, unless this condition is waived in writing by Seller. Proof of neutering shall be sent toSeller.

9. If Purchaser breeds the animal(s) acquired under this Contract Purchaser willaccurately represent the ancestry and percentage of wolf involved to all prospective buyers to the best of Purchaser's ability Seller agrees to provide documentation of lineage and percentage of the animal(s) to Purchaser at time of purchase.

10. Purchaser agrees that if the animal(s) need(s) to be given up for any reason, it shall be returned to Seller. Purchaser will not sell or give the animal(s) purchased under this Contract to another person or entity without prior written permission of Seller.

11. The animal(s) purchased from Seller is/are guaranteed to be free of major congenital defects, such as hip dysplasia. Should any congenital defect appear as the animal matures, Purchaser must provide veterinary documentation to Seller and claims must be made within one (1) year of purchase. Upon documented proof of congenital defect, Seller will accept return of the animal(s) and a full refund of the purchase pricewill be made to Purchaser.Should Purchaser desire to retain the animal, a fifty (50%) percent refund will be made to Purchaser upon receipt of a veterinarian's spay/neuter certificate,and Purchaser shall assume costs of medical treatment.

12. Purchaser agrees to accept full responsibility and liability for the future actions and/or behavior of the animal(s) purchased from Seller, and hereby relinquishes and waives any right to hold Seller responsible or liable for any future actions or behavior of the animal(s) or the consequences thereof.

13. The parties agree that should the animal(s) be confiscated by governmental agencies for any reason, that Seller shall have the first option for retrieval and rescue of such animal(s).

14. Purchaser agrees that failure of Purchaser to comply with the terms of this Contract constitute sufficient cause for repossession of the animal(s) and that any breach of this Contract is punishable under the laws of the State of ______________________.
Any attorney fees or other costs incurred in the enforcement of this Contract shall be assessed to the party against whom judgment is rendered..

IN WITNESS WHEREOF, we hereby afix our names and/or seals this ____________ day of ______________________ ,19 ____.

____
SELLER

____
BUYER

STATE OF ________________)
) ss.
COUNTY OF ______________)

Subscribed, sworn to and acknowledged before me this __________
day of ____ , 19 ___,
by ____(Seller)

____
NOTARY PUBLIC

My commission expires:

____

STATE OF ________________)
) ss
COUNTY OF ______________)

Subscribed, sworn to and acknowledged before me this __________
day of ____, 19 ___.
by ____ (Buyer)

____
NOTARY PUBLIC

My commission expires:

____

CHAPTER III. REGULATIONS WHICH MAY AFFECT YOU

CURRENT STATE REGULATIONS PERTAINING TO WOLVES AND WOLF HYBRIDS COMPILED JUNE 1, 1995

The following is the annual listing of summaries of state Fish and Game agency regulations pertaining to wolves and/or Wolf Hybrids as compiled by The Wildlife Education and Research Foundation and published in *The Wolf Hybrid Times*. The listing is drawn from copies of regulations received directly from the various state Fish and Game Departments.

Please note that these regulations apply at the state level only. In some instances, these regulations are overridden by local, municipal or county regulations. In ALL instances, the Federal regulations regarding possession of pure wolves (i.e., an endangered species) will apply, whether they are actively enforced or not.

Please note particularly the language that some states use regarding "potentially dangerous" or "inherently dangerous" animals. This language leaves the legality of possession open to the discretion and interpretation of the Game Commission or local authorities when such cases are prosecuted.

In all states, whether specifically designated in the state regulations or not, if an animal causes property damage or harm of any sort to a human or another person's animal, the owner is liable for the costs of any damages and possible prosecution by both the victim and legal authorities.

ALABAMA

"Notwithstanding any provision of law to the contrary, it shall be illegal to own, maintain, sell, or trade any canidae or felidae for which there is no USDA licensed rabies vaccine. Anyone currently owning or maintaining such animal may keep the animal for the length of the animal's life providing the animal is spayed or neutered and is registered with the Department of Agriculture and Industries. This section does not apply to any zoological parks, circuses, colleges, and universities, animal refuges approved by the Department of Agriculture and Industries, county or municipal humane shelters, the Department of Conservation and Natural Resources or veterinary clinics."

ALASKA

According to a letter received from the Department of Fish and Game, "Under 5 AAC 92.029 Permit For Possessing Live Game, it is illegal for anyone to possess, import, export, buy, sell or trade a wolf (pure or hybrid) without a special use permit from the Department. The only canid that may legally be possessed without a permit is the dog (*Canis familiarus*). Permits to possess game species are issued only for valid scientific or educational purposes; permits will not be issued for possession of game as pets."

ARIZONA

Wolves may not be possessed in Arizona with out permitand permits are generally only issued for scientific and educational activities or commercial enterprises such as carnivals or fairs. Any animal which is at least 1% domestic dog is considered to be a domestic dog and not regulated at a state level; however, local ordinances may override these regulations. Hybrids may also fall under the jurisdiction of agencies such as county rabies animal control departments.

ARKANSAS

"The Arkansas Game and Fish Commission does not regulate the ownership of pure wolves or Wolf Hybrids."

CALIFORNIA

Ownership of pure wolves is illegal except by the few people qualifying for a valid permit from Fish and Game. Among the criteria for such a permit are rigid requirements for facilities and experience in raising such animals, along with approval of the USDA. First generation Hybrids are considered as wolves. Subsequent generations are considered domestic dogs and are not regulated by the Department; however, certain counties have prohibited their possession. Local regulations may override the state regulations making ownership more restrictive or prohibited in that city or county.

COLORADO

No person shall sell, acquire, purchase, trade or barter live wildlife unless he first obtains a proper license as provided in this chapter ....Except as otherwise prohibited by the Wildlife Commission regulations, a (commercial wildlife park) licensee may buy, sell, trade, barter, propogate, import and export any lawfully acquired wildlife...." The Colorado Division of Wildlife does not regulate ownership of Wolf Hybrids as they are not considered wildlife; they are considered domestic animals.

CONNECTICUT

"Possession of potentially dangerous animals. For the purposes of this section, the following shall be considered as potentially dangerous animals: ... the wolf ... No person shall possess a potentially dangerous animal. Any such animal illegally possessed may be ordered seized and may be disposed of as determined by the Commissioner of Environmental Protection. Any person who violates any provision of this section shall be fined not more than one hundred dollars for each offense...A bird or quadraped which results from the crossbreeding of any animal with one of the species listed above... shall be considered to be a wild bird or quadruped of that species."

DELAWARE

"No person shall bring into this State, possess, sell or exhibit any live wild mammal or hybrid of a wild mammal... not native to or generally found in Delaware without first securing a permit under this chapter." The regulations state: "There must be two enclosures to house a (Wolf Hybrid), a primary enclosure and a secondary enclosure. Fastening or locking devices shall be required on both the primary and the secondary enclosures and must be tamper proof from the general public. The primary enclosure shall consist of a pen, cage or other enclosure where the (Wolf Hybrid) will be kept, and must be of such a construction that it cannot be destroyed by or escaped from by the (Wolf Hybrid). The secondary enclosure must be of a type sufficient to prevent the (Wolf Hybrid) from escaping from the property of the custodian. ...Should the (wolf or Wolf Hybrid) escape from its primary enclosure. The secondary enclosure shall be of a type sufficient to prevent bodily contact between members of the public and the (wolf or Wolf Hybrid)."

"The subject creature must not be a public nuisance. A nuisance will be considered as including, but limited to, a subject creature which creates excessive odors or noise, displays obnoxious behavior or causes justifiable fear."

Inspection of the premises where the animal will be housed is required. Individual animal permits are required for pet owners and Class (4) permits are required for breeders or exhibitors of wolves and Wolf Hybrids.

FLORIDA

"Pure wolves and Hybrids which are 25% or less dog require a $100 permit per year unless exempted as a researcher. Provisions include owner is 16 years or older, has knowledge of the species' requirements and is able to provide a proper diet, health care and exercise, caging meets minimum specifications and that the neighborhood setting is appropriate." There are yard, containment, housing and care requirements. "Those animals that are 75% or more wolf must be permitted as Class II wildlife."

GEORGIA

Possession of wolves and wolf Hybrids is prohibited in the state; "except that any person possessing hybrid crosses between wolves and domestic animals on July 1, 1994 shall have until July 1, 1995 to apply for a fee-exempt permit to possess these animals as pets; provided, however, that the said hybrid is sexually neutered; provided, further, that it shall be unlawful to transfer possession or ownership of said hybrid without prior written approval from the department. Liability insurance shall not be mandatory for wolf hybrids possessed under this fee-exempt permit.

IDAHO	"Any canine exhibiting primary wolf characteristics shall be classified as a wolf for the purpose of identification. All such canines shall be tattooed, registered and licensed by the Department of Fish and Game....Captive wolf licenses are issued at an annual fee of $10.00....Since wolf/dog hybridizations are known to exist within Idaho and these hybrids are not protected by the United States endangered species act, a biological evaluation shall be required of the animal to determine species priority."
ILLINOIS	The Illinois Dangerous Animals Act prohibits ownership or possession of wolves except for zoological parks, federally licensed exhibits, circuses, scientific or educational institutions, research laboratories, veterinary hospitals or animal refuges where they must be in an escape-proof enclosure. The 1994 response included the additional statement: "There is no separate designation for the cross-bred wolf/dog or coyote/dog mix, and as such are accepted as wolves if they are represented as wolf crossbreds." (sic) "It is no defense to a violation of Section 2 that the person violating such section has attempted to domesticate the dangerous animal." Illinois Revised Statutes, Chapter 8, paragraph 242(2).
INDIANA	A Class III wild animal permit is required for each individual wolf Additionally, a USDA permit must be possessed by the owner for each wolf. Wolf Hybrids are not regulated by the state and no longer require a permit. See Indiana Register, Volume 18, No. 6, March 1, 1995, DNR 310IAC 3.1-11-8.

IOWA

Local, municipal or county officials should be contacted for relevant regulations/ordinances which would supersede state regulations. Regard-Regardless of percentage, Wolf Hybrids fall under Chapter 162, Code of Iowa and Iowa Departmental Rules, Chapter 21–67 (which pertains to all dogs). Iowa Department of Agriculture & Land Stewardship issues licenses and facilities must be inspected for compliance with applicable laws before issuance or renewal. Owners of two or more purebred wolves must have a game breeder's permit from the Department of Natural Resources, along with their USDA permit. The state agency also conducts inspections of the facilities. Breeders must have a breeder's license and keep records specified by the Department.

Wolves and Wolf Hybrids would be subject to the same entry requirements as dogs coming into Iowa from elsewhere and must be accompanied by a veterinarian's certificate indicating apparent freedom from or exposure to disease. No animals will be admitted from a rabies quarantined area.

"All animals four months of age and older must have been vaccinated for rabies within the past 12 months, with a vaccine and procedures recommended by the Compendium of Animal Rabies Control. Currently, the Compendium recommends that hybrid animals be considered as wild animals and strongly discourages vaccination. There is no approved vaccine for use in wolves."

KANSAS	"Any person possessing a wolf must have a possession permit. There is no charge for the permit." Proof of purchase or receipt is required. A game breeders permit is required to engage in the business of raising and selling wolves. Wolf Hybrids were not mentioned.
KENTUCKY	"Pursuant to KRS 150.183, wolves cannot be imported or possessed except for certain educational, scientific, exhibition or research purposes approved by the Commissioner. Presently, there are no state regulations that specifically address wolf/dog hybrids; however, this may change in the near future as the state is currently in the process of restructuring pet and propagation regulations."
LOUISIANA	The Louisiana Department of Wildlife and Fisheries prohibits private possession of wolves or Wolf Hybrids. Qualified zoos, educational and scientific organizations are exempted from this prohibition. In addition to the state regulations, there are applicable parish (county) or municipal regulations that in some cases also prohibit possession of wolves or Wolf Hybrids.
MAINE	Wolf Hybrids are considered to be domesticated animals and are not regulated by the Department. However; "Because of the lack of an approved rabies vaccine, if a Hybrid is exposed or is suspected of being exposed, it will be confiscated, euthanized and tested for rabies."
MARYLAND	A person or incorporated or unincorporated organization may not harbor or import wolves... or hybrids, for which there is no U.S.D.A. certified vaccine against rabies....Wolf x dog hybrids are considered to be wolves and therefore are not allowed.

MASSACHUSETTS

"No person shall possess, sell, trade, breed, import, export or release a wild canid hybrid or wild felid hybrid, except as otherwise provided by rules and regulations of the division. Any mammal which is the offspring of the reproduction between any species of wild canid or hybrid wild canid and a domestic dog or hybrid wild canid, or is represented by its owner to be a wolf hybrid, coyote hybrid, coy dog or any other kind of wild canid hybrid... shall be considered to be wild animals and subject to the provisions of this chapter."

"All existing pet hybrids will be 'grandfathered' but they must be registered with the (Department of Fisheries and Wildlife) by July 1, 1994. Although these existing pets are grandfathered, they may not be bred... hybrids obtained after April 10, 1994 are unlawful and subject to the enforcement provisions of the state's wildlife laws (MGL 131)." Applications for the permit require two photographs of the animal to be registered.

MICHIGAN

"Wolves are listed as endangered in Michigan and cannot be possessed, transported, bought or sold without an endangered species permit. Permits... cannot be given for private possession of wolves... Presently, Wolf Hybrids can be possessed in Michigan;" however owners should be cautioned that The Michigan Department of Agriculture will consider Wolf Hybrids that are involved in a bite incident to be unvaccinated and they will be treated as unvaccinated dogs.

MINNESOTA

"Minnesota does not currently regulate Wolf Hybrids, although they may be prohibited by local ordinances in some areas. However, if a

Wolf Hybrid bears close resemblance to a "pure" wolf, the owner may be required to show proof that the animal is legally maintained and that it is a Hybrid. Pure wolves may not be taken from the wild, may only be obtained from properly licensed breeders and require a permit for their possession."

MISSISSIPPI

"The Department of Wildlife, Fisheries and Parks has no regulations regarding the possession of wolves or Wolf Hybrids."

MISSOURI

"Any person holding wildlife in captivity in any manner shall have in his/her possession the prescribed permit or evidence of exemption." Wolves and Wolf Hybrids are considered Class II wildlife and there are minimum containment specifications and recordkeeping requirements.

MONTANA

"Wolves and certain other predators are required to be tattooed and records of the same maintained by the Department of Fish, Wildlife and Parks. If any wolf... that is held in captivity or that escapes from captivity causes any damage to the personal property of another person, compensation for such damage must be paid by the person holding or who held the animal in captivity." The regulations define "wolf" as "any canine which is one-half or more wolf. All 50% or greater crosses and pure wolves are required to be tattooed and registered."

NEBRASKA

The statute prohibits possession of wolves in the state. Wolf Hybrids are not regulated by the Game and Parks Commission. But a Wolf Hybrid which looks like a wolf will be determined to be a wolf.

NEVADA	"The Department of Wildlife does not regulate wolves or Wolf Hybrids. However, other state agencies or local (county/city) entities may have special regulations or ordinances. Wolves are exempt from importation and exportation requirements."
NEW HAMPSHIRE	House Bill 1100 prohibited ownership or possession of Wolf Hybrids effective upon signature (6/6/94). Existing Hybrids are grandfathered in.
NEW JERSEY	"Wolf Hybrids are not regulated in the State of New Jersey. Wolves are regulated as potentially dangerous species and are not allowed to be possessed as pets."
NEW MEXICO	"It is not legal for anyone to take, own or otherwise possess Mexican wolves originating in New Mexico—as these are listed as endangered under the state's Wildlife Conservation Act. In addition, no specimens of this or any other endangered/threatened subspecies of wolf may be brought into New Mexico without appropriate federal permits. All wolves require a permit from the department for importation into this state. Importation permits are issued for scientific research, rehabilitation or temporary display of animals in connection with educational purposes. Wolf Hybrids do not fall under the jurisdiction of the department—given that their ancestry includes a domesticated animal. However, if one of the parents of a hybrid involves a wolf present in New Mexico under illegal circumstances, both the latter and its off-spring are subject to seizure by the department and/or the U.S. Fish and Wildlife Service.

NEW YORK

"No person may possess, release, transport, import or export or cause to be released, transported, imported, except under permit from the department any of the following wildlife...any animal, the overall appearance of which makes it difficult or impossible to distinguish it from a wolf (Canis lupus) or coyote *(Canis latrans).* Such permits may be used only for scientific or exhibitory purposes."

NORTH CAROLINA

"Possession of eastern timber wolves or red wolves (100% purebred stock) would require a wildlife captivity license, which may be obtained by written request to this office for a captivity application and then by following the application procedures. The Commission does not recognize wolf-hybrids as wild animals."

NORTH DAKOTA

Wolves and Wolf Hybrids are classified as Class 5 Nontraditional Livestock by the North Dakota Board of Animal Health. "All non-traditional livestock must be licensed and comply with the administrative rules of the board and applicable statutes." The annual license fee per animal is $10.00, with a maximum fee for persons holding more than one animal, $75.00. Nontraditional livestock may not be released or abandoned without prior written authorization from the board. All nontraditional livestock must be individually identified by tag or identification numbers issued by the board.
"Licensees...shall comply with fencing standards that will assure containment." Fencing must be at least 6' high. "all gates in the perimeter fence must be locked."

OHIO	"The Ohio Division of Wildlife has no regulations concerning the sale or possession of wolves or Wolf Hybrids in Ohio. The U.S. Fish and Wildlife Service has authority on dealing with wolves in Ohio."
OKLAHOMA	"A wolf/dog hybrid is considered a domestic animal and not wildlife You do not require a noncommercial breeder's license to own a single wolf; however, the animal must be purchased from a commercial breeder and a sales receipt must be kept to prove origin. If two or more animals are owned, then a non-commercial breeder's license must be purchased. If two or more animals are owned and bred for resale, then a commercial breeder's license must be purchased."
OREGON	The Oregon Department of Agriculture requires a permit for each species of exotic animals, which includes wolves and wolf crosses. The permit fee is $20.00 for two years and requires the completion of an application form, including the name, address and phone number of the veterinarian who will be treating the animals and inspection of the facilities where they will be kept to insure that they are adequate to keep the animals contained. There are no specific determinations as to size, except that they be humane and adequate and cause the animals the least stress possible.
PENNSYLVANIA	Wolves and Wolf Hybrids may be possessed if they are purchased or received from any lawful sources from within or without the state if the owner first secures an Exotic Wildlife Possession Permit from the Pennsylvania Fishand Game

Commission Inspection by the Commission is required to insure that minimum requirements for secure and humane housing conditions are met. Minimum pen requirementsfor a single animal are "15' length by 8' width by 6' height. For a pair, double the cage length. For each additional animal -after 2 add 10' to the cage length. A secluded den area 4' W x 4' L is required required for a single animal. Add 3' in length for an additional animal."

"Records shall be maintained of acquisitions and disposals of exotic wildlife as well as exotic wildlife born on the premises. Records shall be in ink, written in English and include the full-name and address of the person with whom a transaction is conducted. Records shall be available for inspection by Commission personnel at reasonable hours. Entries shall be made on the day of transaction."

RHODE ISLAND

Wolves and Wolf Hybrids require a permit for importation or possession. Permits must be obtained before the animal(s) enter the state. Pure wolves require a USDA permit. Facilities inspections are mandatory for both. Although it is not specified in the "Rules and Regulations Governing Wild Animal Importation and Transport," the 1994 response included the comment: "In Rhode Island, it has been determined that a wolf is any canine which has 1% wolf blood."

SOUTH CAROLINA

"It is illegal to import a wolf or a Wolf Hybrid that is 25% or more wolf. There is no regulation against possession of a Wolf Hybrid if purchased within the state. Wildlife Commission policy considers them "potentially dangerous". ..." No carnivores which are normally not domesticated may be sold as a pet in this state. Such carnivores shall include animals known to be reservoirs of rabies, such as wolves..."

SOUTH DAKOTA

"A permit from the South Dakota Animal Industry Board is required to import nondomestic animals. In addition, a permit as described in Section 12:68:18:03 is required to possess in South Dakota any nondomestic mammal, or any of its hybrids listed in this section." (All nondomestic members of the family *Canidae*)."

The 1994 response read "Wolves and Wolf Hybrids less than three generations removed from the wild are considered protected animals and may not be possessed. Ownership of "domesticated" wolves permitted provided proof of ownership and purchase is provided. Other Hybrids not regulated."

TENNESSEE

No wolves may be possessed without a permit, which includes very rigid pen specifications. Permits are issued only to zoos, circuses, and commercial propagators. Wolves are considered to be Class I wildlife. Wolf Hybrids of any percentage are not regulated by the Agency.

TEXAS

"No person in Texas may possess a true wolf unless that person obtains and possesses a valid permit from the Texas Parks and Wildlife Department or the United States Department of Agriculture. The permit required by the state agency is $100.00 and is required for each of the first five animals acquired. The permit must be renewed each year, but this law expires August

31, 1997. An additional permit is required under the Texas Endangered Species Act, regardless of which of the above mentioned agencies originally permitted the animal."

Effective September 1, 1995, Chapter 62 of the Parks and Wildlife Code prohibits the hunting, killing or injuring of Wolf Hybrids in controlled hunts, as well as the sale or distribution of whole carcasses or parts of Wolf Hybrids that have been killed in a controlled hunting event.

UTAH

"No permits are required by the Division of Wildlife Resources for Wolf Hybrids. They are considered in the same category as dogs. Rabies shots will not be given to Hybrids by veterinarians in this state because they are not eligible to be vaccinated. Local ordinances may vary from locality to locality."

VERMONT

"It is unlawful for any person to bring into the State of Vermont any live wild bird or animal... unless, upon application in writing therefor, the person obtains a permit to do so." The definition of "Wild Animals" include "the family *Canidae*, any hybrids with domestic dogs." However, ownership or possession Wolf Hybrids is not regulated by the Agency of Natural Resources and are considered domestic dogs, but local ordinances may provide containment requirements. A recent rule (Regulation #1009) defining physical and behavioral characteristics of wolves has been adopted to aid in identification.

The Rule reads in part: "A timely determination on whether an animal is a wolf hybrid or dog is important, especially if one of these animals comes in contact with a human, because there is no approved pre-exposure rabies vaccine for Wolf Hybrids." Wolf Hybrids must be registered, licensed and vaccinated through the municipalities and there is an annual licensing fee.

VIRGINIA	"Pure blooded wolves are not allowed entrance into the state for private ownership. Hybrid wolves are allowed but the owner must obtain a permit from this office."
WASHINGTON	"The wolf, *Canis lupus*, has been classified as an endangered species by the Fish and Wildlife Commission. Private ownership of a wolf requires requires proof of legal acquisition, health cert-i ficate, proper holding facilities. Wolf Hybrids are not classified as wildlife and therefore do not fall within the regulatory authority of the Department of Fish and Wildlife. This does not mean that local jurisdictions may not regulate or prohibit the possession of these animals. Add-itionally, other state or federal agencies may have regulations that apply to the importation or possession of wolves or wolf hybrids."
WEST VIRGINIA	"The importation of any species of wildlife into West Virginia is governed by a Wildlife Importation permit. These permits have been denied in recent years for wolves and other potentially dangerous wildlife due to some bad experiences in the past. Wolf Hybrids with more than 1% dog in their lineage are treated as dogs and therefor would not be under the jurisdiction of the Fish and Game Department; however, importation is not allowed."
WISCONSIN	"Pure wolves are illegal except by permit for zoological, educational or scientific purposes or

propagation for preservation purposes. Hybrids are not regulated.

WYOMING

The regulations received last year stated that: "It is unlawful to import or possess a wolf, wolf/hybrid of ANY percentage in the state of Wyoming." However, the regulations received in April 1995 read: Section 5 (d) "Wildlife Prohibited from Importation. The following wildlife may be possessed without a permit if captured in the State of Wyoming, but may not be imported into the state. (i) Wolves (*Canis lupus*), wolf hybrids, and/or wolf/dog hybrids may not be imported or sold."

Note: These regulations are current to June 1, 1995.

1995 Regulations Pertaining to Wolf Hybrids

June 1, 1995 © 1995

State	Not Regulated	Permit Required	Special Conditions	Prohibited
Alabama				X
Alaska		X		
Arizona	X			
Arkansas	X			
California	X		F1 generations prohibited	
Colorado	X			
Connecticut*				X
Delaware		X		
Florida	X	X	Class II wildlife permit required for 75% or above	
Georgia				X
Idaho		X	Tatoo required	
Illinois				X
Indiana	X			
Iowa	X		Licensed as dogs; facilities inspection for compliance.	
Kansas	X			
Kentucky	X			

State	Not Regulated	Permit Required	Special Conditions	Prohibited
Louisiana				X
Maine	X			
Maryland				X
Massachusetts				X
Michigan	X			
Minnesota	X			
Mississippi	X			
Missouri		X		
Montana		X	Tatoo required for 50% or above	
Nebraska	X			
Nevada	X			
New Hampshire				X
New Jersey	X			
New Mexico	X			
New York				X
North Carolina	X			
North Dakota		X		
Ohio	X			
Oklahoma	X			
Oregon		X		
Pennsylvania		X	Wildlife permit required	
Rhode Island		X		
South Carolina	X		Must be purchased within the state	
South Dakota		X		
Tennessee	X			
Texas	X			
Utah	X			
Virginia		X		
Vermont	X		Importation prohibited	
Washington	X			
West Virginia	X			
Wisconsin	X			
Wyoming				X
				10

The Wildlife Education and Research Foundation

LOCAL ORDINANCES WHICH OVERRIDE STATE REGULATIONS

Many counties and municipalities have ordinances which will override the state regulations. All owners should make themselves aware of the regulations of their own local governments. Listed below are but a few and represent only those of which we are presently aware.

Prescott, Arizona
April 1993

An effort by the Prescott City Council to put harsh and restrictive regulations on Wolf Hybrids met with stiff opposition from local owners and sympathizers. After considering various drafts of permit regulations, the Council finally voted in an animal control ordinance that was much easier to live with. Wolf Hybrids would be treated generally the same as dogs, but when issued a dog tag, it would bear a different color. Rabies vaccinations are required; however, the provision is unclear if veterinarians would be required to give them.

If a Hybrid bites, the magistrate is not required to have the animal killed after the first or second bite; however, it would be left to the discretion of the magistrate and he has the power to order euthanasia.

The only other special requirement for Wolf Hybrids is that they require 6' high fencing placed at least ten feet inside property lines.

Santa Clara County, California
April 1993

Santa Clara County Animal Control has permit regulations for Wolf Hybrids, which specify minimum allowable containment facilities, which in addition to minimum square footage per animal, also require padlocked double gating into the pens, a perimeter fence no less than five feet within the property lines, and vaccination for rabies. However, "any hybrid which bites a human being will be quarantined by the standards required for an unvaccinated animal, even if vaccinated."

"The owner of the animal will provide financial responsibility by posting a bond or certificate of insurance in the amount of one hundred thousand ($100,000) dollars."

"Hybrids must be under the control of a person eighteen years of age or older, and who is physically capable of restraining the animal."

St. Louis County, Missouri

"It is unlawful for anyone to have or bring into St. Louis County any non-domestic animal without first applying for and receiving a license from the Director of the Department of Community Health and Medical Care.

Non-domestic animals include all wolves...or any crossbreeds thereof."

"The law requires you to:

1) Apply for an application for a non-domestic animal license before you purchase the animal;
2) Comply with the 1975 cage size standards of the American Association of Zoological Parks and as specified in the Federal Animal Welfare Act of 1970. Cages of over 80 square feet must be located 100 feet from any property line.
3) Provide liability insurance of $100,000 or a surety bond before the license can be issued. This is required to cover any bodily injuries, deaths or property damages which may result from the ownership of such animals.
4) Obtain inspection and approval of your application from a County Health Officer. If you live in a municipality, you may also have to get a city license.
5) Present to the Director a State of Missouri Wildlife Breeder or Hobby Permit and any Federal Wildlife Permits required for each species before you may be granted a license.
6) Follow the specifications for the humane handling, care, treatment and transportation of non-domestic animals, according to the requirements of the Federal Animal Welfare Act of 1970 and the Wildlife Code of Missouri."

Billings, Montana
February 1993

On February 22, 1993, Wolf Hybrids were officially banned in the city of Billings, Montana. The local ordinance does provide for the grandfathering in of present animals, but with strict regulations.

Animals may not be removed from the kennel without muzzling and leashing, even if riding in a closed vehicle.

Animals must be "kenneled;" however, no guidelines for "kenneling" were provided. "Proper containment" is up to the discretion of the local Animal Control officers.

Wolf Hybrids must be registered as such with the city and must be kept in a "securely enclosed structure to which all entrances remain locked. The structure must be designed so that a person cannot place an extremity in any part of it and so that the wolf/dog hybrid cannot escape." Facilities will be inspected and photos taken of the animals enclosed.

Any animal found not in compliance with these regulations are subject to seizure and euthanasia and the owner will be ordered to appear in City Court to show cause why they should not be destroyed.

The primary reason given by the City Council for the ruling was that there is no proven rabies vaccine for Wolf Hybrids, and said it would have to reconsider the muzzling portion of the ordinance if such a vaccine were approved.

Albuquerque, New Mexico
August 1, 1992

From a pamphlet detailing the Albuquerque Animal Control Ordinance, published 8/1/92.

While wolves/Wolf Hybrids are not banned, they are classified as exotic animals and owners are required to obtain a permit from the Mayor's office.

While all dogs and cats are required to have current rabies vaccinations, "the Mayor may require animals other than dogs and cats to receive animal anti-rabies vaccinations." The ordinance contains no breed specific or species specific language regarding quarantine or euthanasia for testing for rabies.

Another interesting section reads: "No person or household shall own, harbor or keep more than a combined total of four dogs, cats, wolves, wolf hybrids...except no person or household will be issued a permit for more than four wolves, wolf hybrids, coyotes, coyote hybrids or any combination thereof."

New Hanover County, North Carolina
August 1993

Wolf Hybrids are banned in New Hanover County, North Carolina. Onslow County is presently trying to ban them also. Once again, the rabies question is fortifying the County Animal Control agency's argument that they should be prohibited.

In New Hanover County, wolf-dogs were banned under the county's 1987 exotic pet ordinance.

Redmond and Springfield, Oregon
August 1993

Redmond, Oregon recently became the second community in the state to ban Wolf Hybrids. The other is the community of Springfield.

Besides wolves and Wolf Hybrids, the Redmond ordinance also affects most exotic animals except monkeys. Violators can be fined $103 for a Class A civil violation for each day the animal remains in the city. Once cited, owners have ten days to remove the animals before the city seizes them, according to Susan LaChance, the city's animal control officer.

Reasons given for the ordinance are the issues of "dangerous animals" and the lack of an approved rabies vaccine.

San Antonio, Texas
January 1, 1994.

As of January 1, 1994, Wolf Hybrids are not permitted to be owned within the city limits of San Antonio, Texas. The City Code, Chapter 5 includes in its definitions the following language:

> "Prohibited animals shall mean . . . any member of the genus *Canis,* including wolf, hybrid wolf, coyote, jackal or fox, and similar species except Canis familiarus."

Prince William County, Virginia
April 1995

County attorneys have reinterpreted the county's ordinance governing wild and exotic animals and determined that Wolf Hybrids could be prohibited as pets. Existing Hybrids are not grandfathered in and owners were advised that they would either have to move out of the county or surrender their animals for euthanasia.

King County, Washington
October 1994

King County, Washington has passed an ordinance prohibiting ownership or possession of Wolf Hybrids. Hybrids owned prior to passage of the ordinance are grandfathered in. In Seattle, which is part of King County, city ordinance also prohibits possession of Hybrids. Hybrids picked up by animal control will be automatically euthanized.

Washington, D.C.

Wolf Hybrids are prohibited in the District of Columbia.

CHAPTER IV. CONTAINMENT

---

*Specific suggestions for containment facilities and the ways and means ofconstructing them were covered in the book **The Wolf Hybrid** by Dorothy Prendergast, and we suggest you refer to that volume for detail. There are, however, some specific aspects of secure and adequate containment which should be given consideration.*

In general, however, one should always provide strong and secure fencing of at least six feet in height, a large enough space for adult animals to run and play, a good shelter, and of course, adequate food and water. Additionally, if an animal will be alone for long periods of time, a companion animal of comparable size should be provided. This will provide them with constant companionship and reduce the desire to be outside of their own territory.

Carl Miller

I will introduce this chapter with examples of material appearing in newspapers throughout the country. Horror stories that "don't happen to you"—except, maybe! One of the stories involved the death of a two year old child in Michigan. "The mothers were chatting inside the house. They looked out the window and saw the animal shaking something and found the little girl's throat ripped out and dying. The child had played with the same animal on previous occasions and neither mother seemed aware of potential danger."

In a Phoenix newspaper on a weekday, we counted nineteen advertise ments for Wolf Hybrids, most claiming "lovable, dependable," all claiming "socialized," and one claiming "pure Siberian Wolf;" one reading "Loving burglar alarms, family companions, gentle playmates, protector of children, money back guar., $100 hold with deposit." (!) Also, an ad that touted "cubs just back from Alaska being filmed for..."

There is something wrong with the credibility! Long-time Wolf Hybrid owners should be well aware that no matter how reliable your animal's behavior may generally be, they will always suddenly present you with some new, unexplained behavior—whether it be a sudden urge to dig, to steal things from the house, to try to jump a fence (simply to be on the other side of it–or maybe to get into some trouble), when they have never exhibited such behavior before—or, to attack or threaten when they have never before been aggressive in any way.

Wolf Hybrids are large canines and as such, they present a very real danger to children and animals and to adults who transmit their fear to the animals. Prey or dominance behavior is not unique to Wolf Hybrids nor to wolves. I have on many occasions witnessed the same behaviors in German Shepherds, Huskies and Malamutes, Saint Bernards and Great Danes, and I am sure that the same reactions have been observed in other large dog breeds, whether it be exhibited as aggression or in the mounting ritual.

The point is that there is absolutely no guarantee that you and your Wolf Hybrid will never find yourself in headlines—unless some sane, reasonable and responsible controls are exercised.

The Wolf Hybrid Times has been telling us for years that control by state or local authorities is coming, but we don't listen. It has been telling us that there would be two lines of reasoning for such regulations–1) that there are no licensed rabies vaccines for the animals and therefor the animals cannot be recognized as being protected against or spreading the disease; and 2) on the basis of the Dangerous Animal laws.

We didn't listen—our animals couldn't get out, wouldn't hurt anyone. Maybe not—but now we are being faced with a double dilemma. It may not matter any more if your animals can get out and into trouble–they are Wolf Hybrids.

We can either try to lie, claiming they are "mutts" and obtain the rabies vaccine and dog licenses and hope they never cause any problems, not knowing if they are actually protected from rabies, and pay the financial punishment and possible incarceration if they do cause trouble—or we can face the issue head on and come up with good, reasonable containment requirements not only for Wolf Hybrids, but for any large canine.

What is wrong with keeping your animals in your own yard and allowing the other people the freedom of enjoying their own property or public property? What is wrong with requiring owners of every animal to keep their animals securely contained so that they are not destroying other people's property or threatening lives?

There is nothing wrong with it and if we, as Hybrid owners, are to be able to keep our beloved animals, we must wake up to this fact and instead of hiding behind hurt feelings of being singled out, take some affirmative action to protect our animals and our right to have them.

I referred above to "secure containment." "Secure" by any definition means that the animal can not get out and that others (human or animal) can not get in.

There is one recurring theme through most of the scenarios reported by the press in the Wolf Hybrid attacks. The animal is chained. Wolves and Wolf Hybrids should never be chained on a regular basis. Any animal, including a human, will only become bitter, mean and aggressive if it is kept chained. People who do keep animals chained are not only cruel, but are courting disaster.

Fencing is not inexpensive, but the financial (and emotional) risks of not providing it are prohibitive. Consider a $5 million lawsuit for the loss of a child's life. If one is to have a Wolf Hybrid at all (or a large dog of any breed), one should be able to provide room for it to run and play and the kind of fencing that will keep it in its own yard and territory. And since the laws have changed so radically as to be in favor of possible burglars or intruders (especially children), one must be able to reasonably guarantee that he has used every safeguard possible to keep "accidents" from happening. It is a sad commentary that while one must protect oneself from others, one is at the same time punished for doing so. "Beware of Dog" and "No Trespassing" signs must be posted at short intervals all around your property (the distance varies from state to state). In some states, it is even required to post signs listing the specific statutes under which an uninvited person (burglar, murderer or what not) might be prosecuted.

A gate padlock is inexpensive. For those that have secure fencing, this is pretty cheap insurance. A double set of gates is even better.

Regulation of Wolf Hybrids is coming and we, those people who will be directly affected by it, must stop hiding behind our false sense of security and do something now. If we don't, we will probably be facing total prohibition of the animals we love. At this time, the state and local agencies are receptive to adopting reasonable and humane requirements for ownership of Wolf Hybrids. They are not staffed to conduct "wolf hunts" but some states are required to do so anyway.

Instead of trying to fight regulation (which is inevitable), each and every owner should contact the principals of the various organizations striving to prevent prohibitive regulation to make sure that every voice is heard. Prohibitive legislation can not be prevented without your help and any owner who does not meet regulations required by the authorities will have no excuse. Remember, if your help is not forthcoming, the laws are going to be made for you—by someone who has probably never even seen wolves or Wolf Hybrids in person.

"Timber Ghost." Photo: Ellen & Ron Strope

SOBERING THOUGHTS ON WHY GOOD FACILITIES ARE SO NECESSARY

We were all shocked and saddened by the recent surge in severe and sometimes fatal attacks by Wolf Hybrids around the country.

On September 17, 1994, a young Santa Fe, New Mexico boy went into a neighbor's yard where a Wolf Hybrid was chained. Josh Garner was killed and his mother bitten while trying to pull the dog off the boy. Two other dogs were involved and it is speculated that the boy may have gone into the yard to break up a dog fight between the three animals. The Wolf Hybrid was euthanized and tested negative for rabies. The two other dogs were detained.

Several days later, a three year old boy was mauled by a pregnant Wolf Hybrid in Gypsum, Colorado, requiring 52 stitches for bites all over the right side of his body, including neck, face, back, arm and leg. The animal was chained outside an apartment building while the owner went inside. When he returned, he helped the mother pull the dog from the boy.

On October 12th, the four year old Jasmine Peterson was attacked by two Wolf Hybrids in Tucson, Arizona after she fell from a wall into a neighbor's yard. The girl underwent four hours of surgery. The owner had warned the child to get off the wall prior to the accident. The dogs were destroyed.

On October 28th, two year old Josh Worden was visiting his godmother, a longtime owner and breeder in Lucerne Valley, California, when he wandered into the back yard and put his hands into the cage. The boy's right arm was pulled into the cage and severely mauled and his left thumb severed. The cage was surrounded by double fencing except for the one small area to which the boy went. Both animals were destroyed; neither had rabies.

A three year old boy died December 10th after he was attacked by a Wolf Hybrid in the southeastern Vermont community of Townshend. The boy wandered into the yard next to the day care center he was attending where the 1-1/2 year old female Wolf Hybrid was caring for a litter of puppies.

There was no fence between the day care center and the yard where the Wolf Hybrid was kept in a makeshift pen made out of wooden pallets. The boy's death occurred one month after a child was mauled by a Hybrid in Jamaica, Vermont.

By recounting these incidents, we are not trying to sound like the Humane Society of the United States. What we are attempting to do is once again reemphasize the imperativeness of owners taking all precautions and providing secure, escape–proof and entry–proof containment for their animals. And owners must do so before the law requires it or bans ownership of the animals outright.

Let us examine some of the common elements of the incidents cited above. All of the incidents involved young children who gained unsupervised access to the animals without the owners' permission. Most of the incidents involved outright trespass, but this does not absolve the owner of blame. The incidents could have been prevented if there were a secondary, entry-proof perimeter fence surrounding the primary enclosures.

Perimeter fencing serves two purposes: in the event an animal does manage to get through or out of the primary fencing, there is a second fence which, if placed just a short distance, say four feet from the primary fencing, will severely limit its ability to escape from that one. The second and most important function of a perimeter fence is to prevent incidents such as children or adults reaching through the primary fence to try to pet the animals. And, if the perimeter fencing is privacy fencing, it also prevents teasing or exciting of the animals by human or animal intruders.

Many people will feel that perimeter fencing is too expensive, and outside of their means, but how much more expensive would the lawsuit that inevitably follows such incidents be? You would probably stand to lose everything you own, plus be making payments far into the future. Far better to think of it as supplemental insurance and head for the bank.

Some of the photos we receive show animals contained by only four foot fencing. Four feet is not enough! The Hybrid can stand up and look over the fence without stretching a muscle. What makes you think that he couldn't also clear it in a single bound? And although your animal has never shown any inclination to escape and has always been good around children, what makes you think he won't chase that one kid on a bicycle zipping by the fence, like all dogs occasionally do? Wolf Hybrids are not immune from the same behaviors, but they will definitely get all the bad rap for doing the same things that other canines do.

Six to eight foot fencing is essential, for their protection, for your protection, and for the protection of any others who might be tempted onto your property. And we believe that the day is shortly coming when such fencing becomes a mandatory legal requirement for all canines of similar adult size. We've got to wake up and face the facts. All canines are potentially dangerous under certain stimulatory situations. Those situations may differ for each animal. We must prevent them from happening by taking the precautions before they can happen.

And what about chains? Two of the incidents cited above involved animals who were tied. In the first incident, no mention of a fenced yard was made in the reports. The incident occurred in a subdivision which limits the height of fences. But even if there had been a five foot fence, the boy wouldn't have gotten into the yard so easily, and the neighboring dogs certainly wouldn't have been there.

Keeping a dog permanently on a chain is cruel and inhumane in any event, but not only is it cruel, it is downright dangerous! By chaining an animal, one very quickly creates the personality of a convict in isolation. Fear biting, retaliatory behavior and aggressiveness quickly develop, and it is all because of the owner's inability to grasp this concept. Why would one get an animal that is so vibrant and so full of life and affection and then treat it in this manner? Why have the animal at all?

Yet, people do, and these same people regularly complain about the unfairness of laws restricting ownership or requiring certain types of containment facilities. My dog doesn't need it, they say - until the animal ends up being euthanized because of a one-time incident such as those outlined above.

"All bark and no bite" is a classification often assigned to humans, as well as to dogs. Don't believe the words for a minute. There is always that one situation, that one time when a human or animal can be instantaneously stressed beyond the limits of their endurance, and rationale or training are erased. Owners can not afford to take the chance that this one isolated circumstance will never occur.

We have occasionally been criticized for repeatedly emphasizing the importance of secure and adequate containment facilities; however, we feel that we cannot stress the necessity for such fencing enough.

For the protection of your animals and the protection of those around you, you must take steps to prevent any possibility of events such as those cited above from happening, and you must do it NOW, before it is too late.

CONTAINMENT–RESPONSIBILITY

An Editorial

"Dog Attack Sends N.M. Toddler to Hospital," Albuquerque Journal.

"A 2 year old Las Cruces boy was hospitalized in critical condition (4/9/94) after he was attacked by four Rottweiler dogs in a neighbor's yard, Sheriff's deputies said. The boy's mother was also attacked by three of the dogs but was not seriously injured.

The boy was in his own front yard and wandered into the street or near a fence that contained the Rottweilers. The dogs grabbed the child and pulled him into the neighboring yard...County animal control investigator Shirley Wiegel said the dogs' owner will be cited for having unrestrained and vicious animals."

"Dogs mangle 6-year old," Rocky Mountain News (2/13/94)

"Two dogs who escaped from a neighbor's backyard pen in Aliquippa, PA, mauled a 6-year old boy, mangling his head, then attacked police officers who shot and killed both animals. Ronald Lucas was in critical condition Saturday at Children's Hospital of Pittsburgh, which wouldn't release details about the boy. The dogs, a 100-pound Rottweiler and an 80-pound husky mix were being tested for rabies."

"Attack dogs' owners sued; man found guilty in dog attack." Gallup Independent.

"A Belen man was found guilty of violating three animal control laws in connection with an incident last month when two of his dogs attacked a jogging couple. Alfonso and Theresa Jaramillo were confronted by the two Rottweilers while jogging at dawn on January 19th. In a struggle with the dogs, Mr. Jaramillo shot both animals with a handgun, killing one.

Judge Manual Chavez found Zergerly guilty of harboring vicious animals, having vicious animals run at large and not having dog licenses. He also ordered the destruction of the surviving dog.

Alfonso and Theresa Jaramillo filed a personal injury suit in State District Court Wednesday. "This is an issue about people and injury to people," said the attorney representing the Jaramillos. "It's not an issue about dogs. This incident happened because the dog owners were not responsible."

Once again, the editorial column begins with a series of excerpts from newspaper reports. This time, the articles do not affect your animals, the Wolf Hybrids. Or do they? Take a closer look.

The three incidents occurred within a three month period. All of the incidents received only minor coverage in the news and certainly did not command the headlines engendered when a Hybrid is involved. Thankfully, no-one died, so they will not make the bite fatalities listing for 1994. And all of the incidents described Rottweilers. But that is not the point. Actually, it is irrelevant what breed the dogs may have been. Read the last statement of the last article again. The point is once again, responsible ownership. Each and every one of these incidents would not have happened, had the animals been properly contained.

True, two of the incidents involved animals which were fenced—but fenced with what?! Certainly not chainlink. It really strains the imagination to think that a child could be pulled through chainlink. And certainly, if a child could be pulled through the fence, it would be equally as easy for a Rottweiler to push his own way through the mesh.

What of the animals that escaped their pen? Once again, the key words that come into play are "secure and adequate containment"—secure and adequate enough to prevent the animals' escape and secure and adequate enough to prevent trespass by any other person or animal. A responsible owner would have made these provisions long before any sort of incident could happen.

And what of the dogs at large who attacked the joggers? Where was their fencing? Where was their owner? All of the animals were tested and of course, tested negative for rabies. Rabies is not at issue here. Neither is the breed of animal involved nor whether that particular breed is "potentially dangerous" as a whole. The issue is responsible ownership and the provision of secure and adequate containment for the animals. Once again, we are talking about public safety and the responsibility of every citizen to make sure that they do not contribute danger to the general populace. This responsibility includes the understanding that any dog owner is responsible for the actions of his animals and any damages those animals may cause, and it is up to the owner to prevent such incidents before they happen.

So how does this affect you, the people who own the most media-targeted canines in the United States? I'm sure by now that with all the articles and editorials, you already know the answer. If one of your Hybrids were to be in any of the above situations, you know you could expect the following:

1) Immediate nationwide publicity about the incident with continuous follow-up reporting;
2) Instant death of the animal; and
3) Almost certain prosecution and civil lawsuits, no matter what the circumstances.

Even if the incidents were caused by someone else's Wolf Hybrid, you and your animals can expect a strong push by HSUS and similar organizations to ban ownership in your state.

This last is not a pipe dream. It has already happened in several states this year. One incident was all it took to give HSUS the opportunity to lobby and push prohibitive legislation in Massachusetts, Alabama and Georgia, and they are lobbying actively in New Hampshire and Maine. There is even a paid "circuit lobbyist" appearing at legislative meetings in those states. HSUS has formally and publicly stated on quite a few occasions that banning and eliminating Wolf Hybrids throughout the country is one of their high priorities. If one doubts, subscribe to their newsletter, *Shelter Sense* and read it for yourself. If one doubts the large scale of their lobbying efforts, write to the IRS and obtain a copy of their Form 990 filing for the year to see just how much was expended on lobbying. And some of that lobbying was against your right to own your animal.

All of which brings us back to the original point—the responsibility of every owner of whatever breed to keep it securely and adequately contained, to take extra measures to prevent anyone else's access to your animals' yards and to strictly observe leash laws.

As the weather gets warm, it is time to make that (at least) annual inspection of all the fencing, to make sure that it hasn't weakened during the winter, that there are no possible escape routes. If you don't have one, consider putting up a second perimeter fence outside the main compounds, of a sort which would prevent access to the primary fencing by human or canine and which is padlocked at all times. Remember, perimeter fencing serves two purposes: 1) a second barrier to prevent escape, and 2) a barrier to prevent anyone from approaching the primary fencing. Along with spring cleanup, consider installing a double gate entry system, the outside gate being padlocked at all times to prevent unwanted entry. Double gates also prevent unanticipated escapes while you are going through the gates. Each time you go through a gate, make sure it is securely latched, whether you're coming right back out or not. Keeping your animal securely contained at all times is your responsibility, your checkbook, and your animals' lives depend on it.

WOLF HYBRID ENCLOSURES

Vistara Parham

There are two principles to containment. One is physical protection, the other is psychological well-being. For their own safety and protection, Hybrids must not be allowed to free run. A free-running Hybrid can be poisoned, stolen, trapped or shot. It can chase livestock, contract diseases, get into fights and cause endless trouble not only to itself and its owner, but to all wolves and Hybrids. One irresponsible owner can cause untold grief to thousands of owners and Hybrids. Please, protect your animals by containing them, for their protection and for the protection of wolves and Hybrids everywhere.

If a German Shepherd has a fighting or biting incident, people shake their heads, or at worst, rant and rave. If a Wolf Hybrid does the same thing, there is a tendency to ban, exterminate, change laws, call out the National Guard and create a major media event. Our stakes are far higher. No one would ever suggest banning German Shepherds, yet many do suggest banning Hybrids, sometimes successfully. The cause is almost always one irresponsible owner. If you love wolves and Hybrids, contain them properly.

The second principle involves creating an environment which is conducive to the health and happiness of your animals. My pen is 100' by 40' and this to me, would be a minimum. I cannot imagine a smaller space being adequate. The standard "kennel" and "run," I believe, is inadequate and cruel. Wolves and Hybrids need to run, as a bird needs to fly. It is almost as much a physical need as food. The pen must be large enough that they can get up to a full galloping run.

In addition to adequate pen space, our animals also get a sustained run up and down a steep mountain road at 20 mph plus 4–5 times per week.

Having a regular, hard run over 20 mph does wonders for the health and temperament. Hybrids are peaceful and much calmer when properly exercised.

Leash exercise, that is going for a walk with the owner on a leash, is a great way to spend time together and I highly recommend it (I do it nearly daily myself), but it is not to be considered as exercise for the Hybrid. Our best pace, even jogging, is a pitiful turtle crawl to them.

Different methods work for different people. One man in San Francisco runs his Hybrid many miles per week (on a leash) with his bicycle, and swears by it. His 58% female was hyper and nervous. When he started her on a regular intense daily exercise program, she relaxed into a far more peaceful and contented personality. There are a number of commercial businesses selling specially designed equipment for bicycles which keep the dog away from the wheels of the bike, but prevent him from moving more than two foot away from it. Naturally, one would not want to do this where there is traffic. Other people work their animals by weight pulling, sledding, or with a 3-wheeler, etc.

My point is that strenuous, air-gulping exercise that fully expands the lungs is critical to the health and well-being of Hybrids and the type of containment you provide can go a long way towards fulfilling that need. It must be large enough for jumping, galloping, running play, and secondly, a companion is necessary.

Wolves are social creatures and will never be happy alone. Two (or more) together is a far more natural and happy situation, plus they will run and chase and wrestle, getting far more exercise than they ever would alone.

Within the pen, there must be a house or shelter for each animal which remains dry inside under all weather conditions. The house should be 4' x 4' by eight feet long with a 2' x 2' opening (or larger). There must be ample shade, preferably several areas. A nice shade tree or lots of bushes is great.

The containment area must be kept clean, droppings removed, and fresh water provided at all times. In cold climates, frozen water should be replaced with warm water twice daily.

Hybrids like a good view. Ours love to lounge on top of their house, sun themselves and survey their vast domain. They want to see and watch everything for miles around. Remember, place their house in the middle of the pen, not near the edge where they can take a flying leap and jump out.

They also like their quarters where they can see your house and observe all the arrivals and departures of the people. They want to scrutinize all activities. When my male sees something unusual, he will bark a certain bark to let me know I should come take a look. When I respond to his summons and he sees my face in the window, he gets that "dignified satisfaction look," pleased with himself that he has performed his important duty. Before looking, I can tell by the sound of his bark if he is informing me about a strange vehicle, unknown persons, or an animal in the vicinity. I know that the visual contact between our house and theirs is very important to them.

I like to place objects in the pen for them to examine, play with or tear apart. One of the most popular items is a bale of straw. Recently, I put an old church pew up there and they love bouncing off it, running over and under it, stretching out on it, trying to pull it over, etc. It's also great for me–I can take a cup of coffee out there and sit down. Last year, we dragged in a huge dead stump with roots and that provided endless hours of pulling and chewing. Now it serves as a shade space and ambush spot. They get so creative with so little. I think it's a good idea to rotate objects (toys) provided for them. They need stimulation and changes in their yards. Last summer, we put in a child's wading pool and filled it with water on hot days. That was a great hit!

Physical containment is essential for the protection of our animals, and psychological well-being, including adequate space, companionship, exercise, cleanliness, a cozy house and closeness to owners, is just as important to the happiness and well-being of our beloved critters.

SECURITY CAN BE BEAUTIFUL

Your Animals' Compounds Don't Have To Look Like A Prison

Mark Martin

Spring is here, and the annual fervor to clean up, fix up and green up is upon us. After months of dreaming over garden catalogues, we set out determinedly to maintain, beautify and improve our own little corner of the earth. We look at the splendid landscaping in the magazines, weep because major areas can't be improved or blend in with the landscape because of our compounds for our Wolf Hybrids, and worry that our property values may be endangered because our yards look like prison compounds. We seek to provide more natural landscapes for our animals while at the same time, providing locked, escape-proof compounds and compatibility with the environment in which we live.

Good, secure containment facilities do not have to be unsightly or resemble a prison compound. In fact, they can provide a very good vehicle for improving the overall visual appearance of your property. But let's start from the beginning, within the animals' enclosures.

Every property and every climate is different and every property

Beginning to modify terrain with no heavy equipment

owner's perception of "beautiful" is different. Let's say you have about a quarter acre of land to play with. You have chosen to fence that quarter acre with a good quality chainlink fencing to contain your animals. But it looks like a blasted prison. The ground is flat, and to you, it just looks like an exercise yard at the jail.

Let's look again. There is nothing to say that the land has to stay flat. With just a shovel and perhaps a little added dirt, you can rearrange that area's surface to create little "hills and valleys" to obtain a more irregular landscape. Just remember to keep the natural drainage aimed away from the house. You don't want your basement or crawl space to be flooded during monsoon or melt-off seasons!

Large tree stumps can be "planted"—stumps which are cut off at a

height that the animals can jump up and perch on, or small stained and waterproofed bridges over the little gullies you might cut. You might want to cut these gullies as irrigation ditches to water the trees which you may have on the property. With the proper grade, water would only have to be introduced at one end of the ditch until it was full and all the trees had received this necessary moisture. Of course, if it rained, you wouldn't have to irrigate at all. The irrigation ditches really shouldn't be directly along the fence lines, or even directly underneath the trees. They just have to run within the boundaries of the root structure of each tree so that they can utilize the moisture. Irrigation is far superior to sprinkling anyway, as it provides a deep-reaching soak for the trees.

Some owners like to add pools for the animals to play in. Most of us have learned the hard way that air-filled plastic pools may as well be forgotten. If they aren't chewed up immediately, the air will be let out very quickly as toe nails pierce the soft plastic. Even the stiffer plastic pools usually do not last long unless they are sunk into the ground and the top edges protected from curious teeth. Still, even this alternative has its drawbacks. You will still probably have to replace the pools annually as the plastic deteriorates from the weather. Worse, cleaning the sunken pools can be a real problem as the mud and silt from the surrounding area collects on the bottom of the pool. You will also have the problem of how to empty it so that it can be cleaned. Most sunken pools will require purchasing a small sump pump. Such a pump can be obtained rather inexpensively from places like Sears and can be used with an ordinary garden hose.

I personally favor using the fairly indestructible round stock tanks which can be found at most feed stores. They have a drainage valve at the bottom of the tank, to which can be fitted a garden hose and the water gravity-fed to any location you wish. If you simply place the tank on a flat surface on the ground and surround it by a small deck of the same height, the carrying of mud and silt into the "pool" will be reduced, while at the

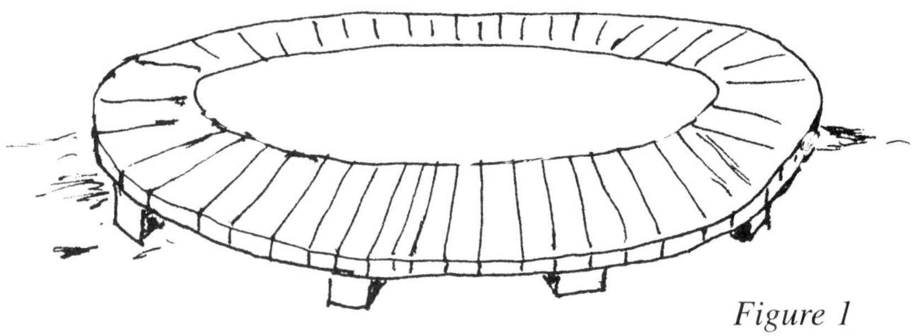

Figure 1

same time, camouflaging the tank and adding another attractive area to your landscape.

Build the deck out of 2" x 6" pine secured to a sturdy structural framework. Many different designs can be used, but probably angling the boards out from the tank and then trimming them in a circular fashion and rounding off all the sharp edges with a router will be the cheapest and easiest to accomplish. Naturally, you will want to stain and water seal the wood deck every so often to keep it attractive and waterproof. (See Figure 1).

We need a shelter or two for the animals—a dog house or two, but they really don't have to look like one. What if it looked like one of those neat

redwood covered homes up in the mountains or a log cabin instead? What if it were water and weather-proof and attractive at the same time?

An attractive shelter can be made rather inexpensively—one which will blend in with almost any environment.

Assuming two large canines using such a shelter, a roughly 4' x 8' shelter is adequate and perfect for most cuts of lumber and materials. Let's build a framing just a few inches short of those dimensions from pine 2" x 4"s for framing. Onto this framing, we nail 1/2" outdoor plywood and cover it with roofing felt (the top being a full 4' x 8' sheet of plywood so that there is an overhang.) Then on the sides and the top, we secure 2" x 6" pine planks placed tightly together and use a router to round all the edges so that there are no sharp edges or corners that can injure the animals. The rounded edges also make it more attractive. We then seal the exterior with a waterproofing that contains a stain which can either make them look like redwood or blend in with the natural colors of the environment. Naturally, you will want all your wooden additions to be of uniform color and design.

The shelters don't have to be high, just high enough in the interior that the animals can easily turn around when they are standing. And they don't need a floor. You'll find that they'll generally prefer to dig their own little indentations or dens in the dirt anyway. And actually, they may rarely use them except in hot weather or to use as a perch. (Hybrids love to play "king of the mountain.")

But what about water running in and flooding them? This can be easily avoided by using the same principles as homebuilders and campers. Dig a ditch about one foot out from and around the shelters. Fill it with loose stone so that the water can run through it. Naturally this "ditch" will have to have somewhere to drain, so make sure it can drain away from the area of the shelter. Then take the remaining dirt you initially dug from the ditch and slope it back up onto the sides of the shelter. This dirt covering will not only help keep water from seeping in, but will provide additional insulation from the elements.

While we are beautifying, here's a suggestion for those who normally provide their Hybrids' water in a bucket each day. Go to a nursery or chain store which has a garden department and buy a wooden planter that your bucket will fit into somewhat snugly. When you change the water, you simply have to take the bucket out, empty and refill it and place it back in the planter. Bingo! You have an attractive and inexpensive camouflage for the metal bucket in your new habitat.

Now we have the fence, the uneven terrain and the shelter. But it still looks barren and yuckie out there. This is where you have to make a judgment call. What sort of natural vegetation grows where you live? What sorts of "weeds" have withstood all the pounding, chewing and destruction they have received (and are not poisonous) and still keep coming back—strongly! Consider encouraging those weeds, at least in certain areas of the compound. They are green, they can be pretty, and you don't have to let them spread anywhere else.

Please don't misunderstand me. I am not saying that people in the south and west should encourage cactus or thistle bushes, and I'm not saying that you should encourage weeds that are particularly objectionable to you; but there is other vegetation which can endure considerable destruction and come back with renewed vigor which, if controlled, can improve the natural image in your compounds without threatening the animals. There are lots of natural wild grasses and shrubs which not only blend into the environmental setting, but are super-sturdy besides. Buffalo grass is great for areas where there is little moisture. The grasses and shrubs not only improve the visual image but in nature's way, improve the quality of the terrain where they grow.

Do not expect to have a perfectly manicured yard. With Hybrids, it is not only unlikely, but rather unrealistic. Besides, mowing will be a problem because of all the nice little renovations the Hybrids are constantly making to their yards. For example—Look at that grasshopper! Pounce! Where'd it go? DIG IT UP!!"

Let's move back to the fencing, the visual image that is received by

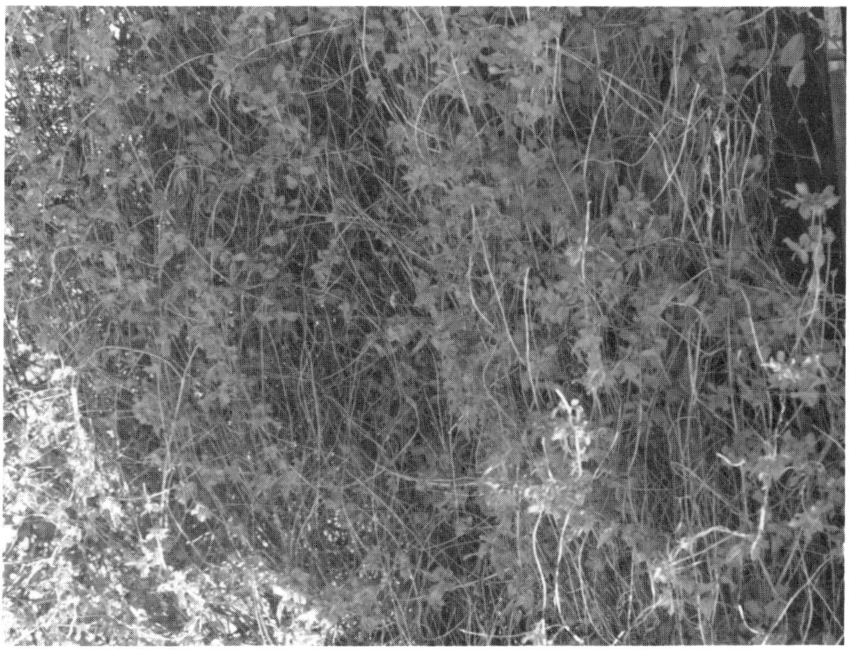

people outside the fence, and to the security of the compounds themselves.

Good, secure containment facilities do not have to be unsightly or resemble a prison compound. Chainlink makes a great trellis for vines, and a heavy vine cover not only deters the curious from trying to peep or reach in, but seriously restricts the curiosity of the contained animals about the world outside, unless someone is right at the fence talking to them. And the fast-growing vines are beautiful additions to your landscaping.

Depending on the climate in which you live, there are many fast growing vines that can be planted on the outside of the fence. Silver Lace vines are probably the fastest and densest growing vines for all areas. Trumpet Vine, Ivies, Climbing Honeysuckle, climbing roses, Bittersweet and Wisteria are good also, but not as fast growing and some of these vines may not be suited for your climate. Suddenly, as the vines start growing, instead of a compound, you have a lovely landscaped yard shrouded by beautiful plantings.

Another option is to plant dense, fast growing hedgerows along the outside of the compound fence. These can be plantings of privet hedge,

A view from the inside of the compound.

Chinese elm hedge, honeysuckle bush or juniper and the like, depending on the climate in which you live. Kept pruned, these hedgerows not only provide a protective and decorative camouflage, but also a handy windbreak. In warmer areas with higher water tables, this same sort of protective shrouding can be accomplished by bamboo or cat tails which will form into a very dense and natural barrier around the fences.

Let's go back into the compound again. All the way around the vertical fence in the compound, you should lay "ground wire" and tie it in securely to the vertical fencing every three inches or so. This is to prevent the animals from digging out. The ground wire can simply be sturdy field fencing which is about 4' in height (if stood vertically). But here is all this ugly wire fencing showing!

ALFALFA–NOT ONLY FOR HORSES

Once the ground wire is in place and secured both to the chainlink and pinned to the ground, a permanent, durable planting of alfalfa will not only mask the ground wire, but improve the quality of the soil. Go back to the farm feed store and ask for alfalfa seed. Now, there are several kinds You want the kind that produces the lavender-colored flowers—perennial alfalfa. Scatter it evenly by hand over the entire area covered by the ground wire and keep the ground damp. From its first sprouting, the alfalfa will produce soft, bushy clumps which, as they grow, will fill into a lovely mass planting and appear to be a short hedge. If unpruned, it will grow as high as two feet and when the lovely lilac-colored blossoms begin to fade, the plants should be trimmed back to within six inches of the ground to keep it from getting leggy and to encourage new growth. These lush green perennial plantings will please your eye from very early spring until late fall, and when they die back over the winter, you will be amazed at how much of your ground wire seems to have disappeared into the soil.

The animals cannot kill the alfalfa, either by chewing or digging at it, once the plants have established themselves. Given at least infrequent watering, they will renew themselves continuously, and at the same time, add nitrogen and other nutrients to your soil.

TREES AND HYBRIDS

You are now asking, "what about the trees? The animals need shade." Trees and Hybrids pose a special problem. No tree is safe from a determined canine that fixes its attention on a certain stick (or tree trunk, for that matter) unless precautions are taken initially. New plantings must definitely be fenced or you will find that still fragile 3 year old tree laying on the ground in pieces the same day you plant it. Even older, sturdy trees must be protected.

I rather like the option taken by the Rio Grande Zoo to protect the older trees that are in the wolf pens. They have taken 2" x 4" pine boards, stained and waterproofed them, and then bound them around the base of the trees in a circular fashion with strips of metal (similar to that used to secure packing crates) to keep those massive claws and teeth away from the tree. These protective casings are as high as the animals can reach when standing on their hind legs. They blend in so well with the environment that visitors rarely even notice them, and they do save the trees! (See Figure 2).

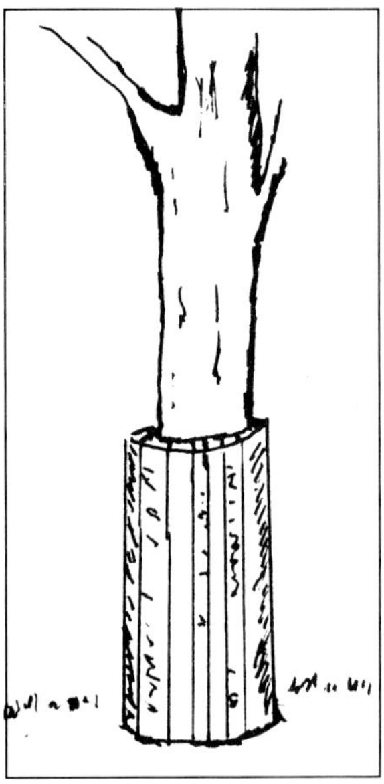

Figure 2

BEAUTIFICATION FOR YOU

Within the compound, we now have trees, an attractive, uneven terrain, attractive "dog houses," an attractive and functional "pool" arrangement and a permanent drainage/irrigation system. The only thing lacking is perhaps a deck for the owner.

Home-built decks need not be expensive and they need not be built out of redwood. Beautiful decks made from pine which is stained and waterproofed serve the same purpose and help save our few remaining redwood trees.

There are many "how to" books available in both lumber yards and libraries which give dimensions and designs for building such decks. They are a lovely addition to any home, and will give hours of pleasure watching your Hybrids enjoying their newly created habitat.

PERIMETER SECURITY

Let's move back outside the compound again. The fencing is secure and shrouded by lovely foliage and it no longer looks like a prison compound. But are you and your animals safe from intentional or unintentional trespass by children, adults or other animals? Not if we listen at all to the continual bombardment of reports of bite incidents and tragedies where children just wander into neighboring yards. We're missing a perimeter fence and we're missing the "No Trespassing" and "Beware of Dog" signs. The signs are easy, but in many neighborhoods, the perimeter fence may be a problem.

While the optimum would be to erect 6' cedar privacy fencing all along the property lines—one which has a locking gate—some localities will not allow you to do this and many cannot afford this more luxurious fencing or do not want to totally block off their view of the world. There is a simpler, less obtrusive and much less expensive alternative.

Not less than four feet from the outside of the primary compound fence, "plant" and waterproof uniform cedar fence posts. These do not need to be closer than 10' of each other. To these posts, you will attach screw or nail mount insulators to a height of 4' and spaced 6"–10" apart on the posts. Now you are ready to string the thin wire and mount the charger for your electrical fence system.

Now, keep in mind that this spacing is optimum. Normally, a four-strand electrical fence is more than sufficient to keep a curious trespasser from trying to enter the "no man's land" between the two fences. I can pretty much guarantee that no child or adult is going to persist in his efforts to climb over or through it so long as the electrical current is going through.

When you buy such a system, read the manufacturer's specifications carefully. Some units are designed for a maximum charge for 1/2 mile; some for one mile, and others provide varying degrees of charge. Remember that the more wire footage being powered by one charger, the less the charge will be toward the end of the fencing. You do not want to use one designed for 1 mile to surround 1/2 acre–the jolt would be too strong and you do not want to injure an unintentional trespasser. Figure out the distance your wiring needs to cover and purchase your unit according to those measurements. Additionally, you should post a sign in the areas where people would be most likely to approach, warning "Danger—electric fence."

There are several types of electric fencing units available. There are those which can be plugged directly into household current. There are those which operate on common replaceable batteries and there are those which operate on solar powered batteries, which are more expensive and require a separate charging unit. The advantage of those which operate directly on household current is that they are less expensive in the long run because of not having to replace batteries as they wear out. On the other hand, if your household electricity is interrupted, the battery operated fencing will still continue to do its job.

We now have a safe, secure and attractive habitat. Given a year or two for the growth of the foliage, your habitat could qualify for a landscaping award. Some of the basics of landscaping are uniformity of color and design in the man-made structures, choosing those colors which best blend with the environment, and choosing plantings and designs which best blend in with the area in which you live.

Creating an adequate, secure and attractive habitat for your animals need not be expensive, and it need not all be done in one year. Security - yes!—it must be provided immediately. But the beauty can develop as you are able to accomplish it and is only limited by nature and your creativity. Happy planning!

CONTAINMENT SURVEY RESULTS

In 1990, *The Wolf Hybrid Times* conducted a survey of subscribers regarding the containment, care and feeding of their Wolf Hybrids. The following summarizes the responses received with regard to containment facilities.

"Shiloh." Photo: Jane Anne Hodges

It is always difficult to convert a survey to statistics, no matter how specific the questions may be. Many answers overlapped several questions. Of 103 completed questionnaires received, most of these we suspect were from the owners most satisfied with the facilities they have. Some sent photographs of their facilities, some of which are reprinted here. Based on those 103 returns, the following summarizes the responses to the survey.

Slightly more than 50% of the owners reported having 50%–75% animals, but many of these owners also had higher percentage animals along with them.

Those reported were usually one or more years old; however four percent of the responses came from owners of one Hybrid of less than one year. 85% of the owners reported that their animals were more than two years old. 36% were four years or older.

Some reported pure wolves, but all of those owners also had other Hybrids.

NUMBERS OWNED

32% had one animal
28% had two animals
26% had 3–4 animals
8% had 5–7 animals
6% had 8+ animals

FENCE HEIGHTS were reported as:

4 foot–8%
5 foot –12%
6 foot–56%
7 foot–1%
8 foot –12%
10 foot–1 response

ANIMALS PER PEN

5% reported two to a "pen." The pens' size for two only, "statistically" averaged 2,112 sq. ft. per pen (roughly 40' x 50'); however, in reality, most of these pens were approximately 400–500 sq. ft. (40' x 10') or less. The reason the "statistical" figure is so inflated is that some subsidiary "exercise areas" were included in the responses. Where dimensions were given, "holding pens" or "kennels" were between 8' x 10' and 8' x 16', although we received one answer specifying 4' x 4'.

"Running areas" ranged from 400 sq. ft. (10' x 40') to 25,600 sq. ft., with the average being around 1,500 sq. ft., no matter how many animals.

48% of those responding used chainlink; 10% used wood board fencing. Others described their fencing as "brick, cyclone, hurricane or block fence."

21% used "tilt wire" or overhang security wire.

30% reported using ground wire, 13% concrete or cinder block, and 6% using rock piled by the fence or stepping stones to prevent digging.

29% reported using electric wire, 6% top only, 10% top and bottom. 13% bottom only about 1' out from the vertical fence; 3 people reported using 3-strand electric fence in conjunction with their other fences.

17% had totally covered pens. 29% reported perimeter fence surrounding all other pens or yards.

97% reported locking or "lockable" latches and/or padlocks of some sort; however, only 68% indicated that the gates were actually locked, even "most of the time." "Locks" ranged from clips to U-bolts to padlocks and combination locks. 14% said they never locked them.

Only two responses indicated walkways separating pens. (See individual comments of owners).

In response to the question regarding tying or chaining animals, 59% reported never tying their animals; 21% on occasion, 5% "when I'm at work," and 8% reported tying their animals permanently.

As to whether the animals were allowed in the house, 71% reported that their animals were allowed in the house—most whenever the animals pleased. 10% reported their animals were never allowed in the house. The rest responded "on occasion" or "sometimes."

Other "special precautionary measures" ranged from "Beware of Dog" signs to flood lights at night or perimeter lighting

One reported having the latches and locks 7' above ground level so that children could not reach them.

INDIVIDUAL COMMENTS OF THE RESPONDENTS

"Supervision and proper containment are the key. Not letting dangerous situations occur usually prevents any problems. You must provide as good or better conditions than your local zoo for these animals."

"My Hybrids don't seem to mind being tied in their runs, but when I am able to let them run free in the yard or in the woods where I walk, they are always on their best behavior."

"Yard is escape-proof by animal and entry-proof by humans. We also have XL crates in all bedrooms for extreme weather conditions, hot or cold. Animals spend time in the house but are always supervised. One pair is frequently taken for walks in the desert with no problems. We also have an extra pen on the opposite side of house used to separate animals at breeding and birthing times, same specifications as other pen."

"My biggest problem is with people feeding him and children teasing him. Therefore, I will install a second perimeter fencing to keep people away from him when I'm not there to supervise."

"We believe that responsibility of owners must begin with the proper facilities to protect their animals from harm. People with "make-shift" pens are causing the problems for everyone else's right to own these beautiful canines."

"When children come into the yard, the dogs go into the kennel and children are supervised and kept away from the dogs. The dogs don't like little children anymore."

"Both were raised in the house when young, but preferred to move outdoors. The cost of the enclosures was well over $2,000, but it's worth it, especially as we keep Wolf Hybrids across the street from our neighbor's sheep!"

"I feel all owners should be encouraged to have walkways between their pens to prevent the wolves from injuring each other."

"I believe all Hybrids love and need a high spot where they can get a good view. These are their favorite areas. Hybrids/wolves need to see far —so I think any Hybrid pen should provide an elevated spot where they can lounge. If they are closed in and can't see, it will make them uneasy, uncomfortable."

Photo: Janna Camillo

Photo: Lowell and Linda McGinnis

"Lone Wolf's Utah Of the Wild Bunch." Photo: Jean McCarthy

CHAPTER V. TRAINING

TEACHER OR STUDENT

Bobbie Holaday

I am just finishing what might be called an exercise in frustration, having taken my Wolf Hybrid "Luke" through the novice and intermediate training classes of a regular dog trainer. She is reputed to be one of the best in the area and is very thorough. Many of her own dogs and the dogs attending her classes have won in dog show competitions. But, as has been pointed out in various articles on the Wolf Hybrid, you just can't expect to train an animal with a high content of wolf blood to perform like a dog.

I'd like to point out several contradictions between what I was taught in these classes as opposed to what I've learned from *The Wolf Hybrid* book, an article by Mary McBee and Carl Koehler's obedience session at a recent High Mountain Rendezvous.

Many of the figures and finishing touches taught in the obedience class seem a little ridiculous for a Wolf Hybrid. I refer to such things as the "handler's finish", the "figure eight" and so forth. I much prefer to spend my training time on getting Luke to sit, stand, stay and heel well than on cutesy little polishing tricks that probably impress dog show judges but don't do much for you around the yard or out in the woods. Certainly, off-lead training is excellent and takes many hours of time. Unfortunately, if not started at a very young age, it is particularly difficult.

In the matter of feeding, my instructor advised me to use only a good kibbled food with only an occasional "treat" of canned or cooked meat. Furthermore, she advised me to use a low protein content kibble because the high protein feeds provide animals with too much spirit and make them harder to train. She told me that raw meat was very bad and could interfere with his digestive system. This certainly doesn't agree with what I've learned from Wolf Hybrid sources. In fact, since using a high protein diet, including one meal of raw beef kidney, chicken livers or gizzards or the equivalent a day, I find that Luke produces a much firmer stool, has less gas, and his performance, while very spirited, is strong and consistent.

In the matter of affection, my instructor advised me to give Luke verbal praise for good performance, but not to overdo the affection or he would tend to become too dominant. I find that the more affection and attention that Luke gets, the more cooperative he is inclined to be. It is when I am extremely busy and don't provide him with lots of love and play time that he is unpredictable and uncooperative.

On the subject of mouthing, I noticed that since Luke has been a small pup he enjoys gentle mouthing on my arm. Since this appeared to me to be a form of affection and he never offered to bite, I allowed this for some time. My instructor advised me never under any circumstances to allow such mouthing because it was not a form of affection but rather a show of dominance that could lead to eventual biting. On this one, I am placed on the horns of a dilemma, since Luke acts rejected if I punish him for mouthing, while my instructor keeps telling me I should "bust him in the mouth" for his terrible habit.

During the periods when I am working him, I do not allow the mouthing, but in the early morning when he greets me or when we sit together on the old couch that I allow him to share with me, I tend to allow mouthing when it is gentle.

Frankly, I don't think that "dog trainers" should include Wolf Hybrids in their classes unless they are familiar with this animal. In that case, they should hold "Wolf Hybrid Only" classes and teach obedience, behavior and feeding accordingly. I have found that some of the other handlers are afraid of Luke and of course, he can't understand this because he is just one big lovable, playful guy. Furthermore, the instructor obviously doesn't like him very much and frequently refers to him as "that animal." She often recommends that I keep him apart from the other dogs so that they won't have to confront him. Sometimes I go home feeling like an outcast!

Editor's Note: *Training of Wolf Hybrids is best done at home and started early. As you will gather from the following, obedience training is as much a matter of training the owner as training the Hybrid.*

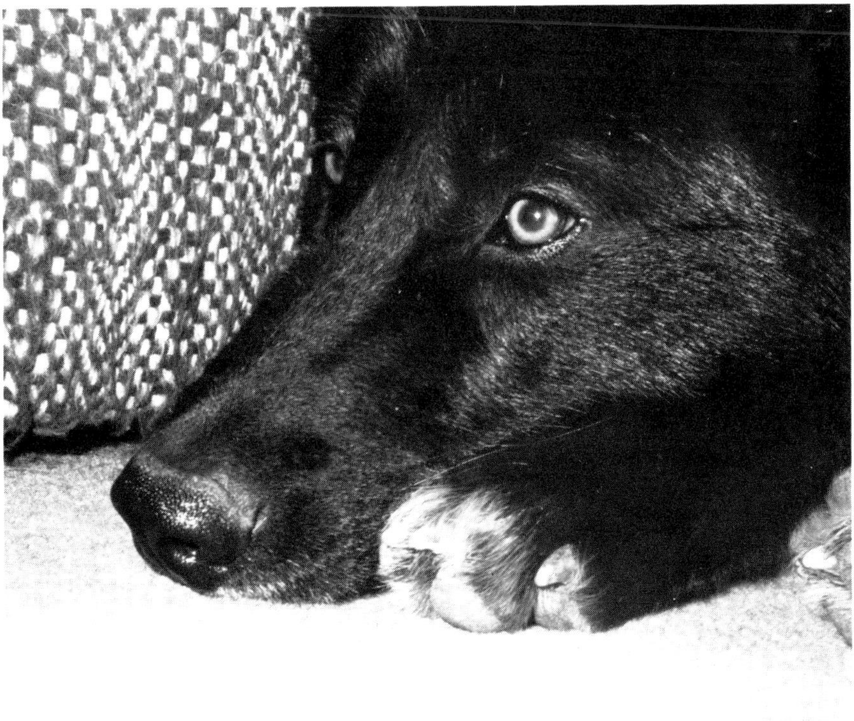

"Denali." Photo: Darlene Sparks

OBEDIENCE TRAINING BEGINS AT BIRTH

D.J. Stover

Leash and obedience training—what's the difference? Most people equate it as one and the same. But obedience training begins long before the leash training part of it. Obedience training begins with choosing the words and teaching the puppy what they mean. For example, we start from the time the puppies begin stealing their first semi-solid food from their mother's bowl.

Each time we go out to feed, we mimic the mother, using a soft but high pitched voice, calling "kid, kid, kid, kid"—a sound which the puppies very quickly associate with food, and come running out of the den or shelter when they hear it. This is the puppies' first real understanding of and response to human commands. We use this form of calling to come to us at our kennels all through the animals' lives, rather than the standard "come" command. Because it is an ingrained signal to them and is associated with a pleasant reward, the animals are much more responsive to it than a stern "come," which to them sounds like a reprimand. Make your call sound cheery, even perhaps silly. Our motto here is "if it works, use it."

We use this call at least twice a day—each time we go into the pen to feed, clean, or just simply check on and play with the puppies. At first the mother stands by somewhat perplexed as the puppies stagger excitedly toward the call. But as she too responds to the call, the puppies trustingly follow her lead, and as their physical agility develops, soon come running at any time of the day, straying farther and farther from their mother and the den.

As we sit and handle and play with the puppies, we gain their confidence and that of the mother. At first, she is cautious and uncertain if her babies are ready to be handled yet, but as no harm comes to them and the puppies lose their fear, she begins to take longer and longer sojourns from the pen, though checking back frequently to make sure all is well.

Up to this point, we have been feeding the puppies within the confines of their shelter, allowing the mother to do the clean up and checking of the puppies after they have eaten. When the puppies also begin to venture further and further from the security of their den, we begin another phase of their training—our version of the beginnings of housebreaking.

The food bowls are now placed some distance outside of the shelter and the puppies called to that spot. Don't worry, they'll still come bounding out to you—after all, you've got the food. The mother still does the job of cleaning the pasty gruel off of them when they have emptied the bowl, but now, because they can only go so far before they have to eliminate, they will slowly and subtly begin associating the great outdoors as the place to do so. Now when the mother does allow them to nurse, she also will quickly exit the shelter when finished, and the pups scramble after her, doing their elimination outside and freeing her of having to clean up after them.

As the pups become stronger and are following their mother through the yard, we will follow through on this initial conditioning by using the mother as an example.

The feeding area is moved into the kitchen of the house, with its easily cleaned vinyl floor. The door is left open so that the puppies can come and go without problem. As the mother and the puppies finish their bowls of food, the mother promptly heads out the door to relieve herself, and the puppies in their rush to follow their mother, do the same. At the same time, we vocally encourage them with cheerful urgings to "go outside, go outside."

Gradually, after days of repetition, the puppies begin to associate and understand the word "outside" and when they are in the house sleeping or playing, we watch them closely for actions indicating they have to urinate or defecate. When we see such actions, we quickly but gently snatch them up and take them outside, cheerfully repeating the words over and over again—go outside."

Never, ever scold or punish them for accidents that occur in the house. Remember, their little bladders and gastro-intestinal systems are still as yet not fully developed, and they do not have control over these body functions as yet. The puppies are barely six weeks old. Each time they wake, each time they eat and each time they drink, they will have to eliminate within seconds and if they are conditioned by following their mother outside from early on, they will begin to associate both the action and the words with the idea that "outside" is the place to go to eliminate.

At this point, through association, you are subtly teaching an action that the body will gradually in the next couple of weeks, be able to conform to. By utilizing the mother's own methods of training her puppies and encouraging them to follow her watchful example, we are usually able to fully house train our puppies by eight weeks of age. Of course, there are the occasional accidents, but by making sure that they go outside immediately when they wake up or after they eat or drink, we are gently and subtly teaching them desirable behaviors in a manner which they can understand. When an owner has trouble "housebreaking" his puppy, the problem is usually the owner's inattentiveness or his lack of immediate response to the postures that signal the need to eliminate. Remember that for the first 10 to 15 weeks of its life, his physical responses are immediate and he cannot "wait a minute" till you finish what you are doing. And again, remember that when a puppy wakes from a nap or finishes his food, he can only go a very limited distance before his bladder takes control and an "accident" occurs. It is not disobedience.

As the litter reaches 6 to 8 weeks of age, the puppies are tumbling playfully over each other, pouncing on sticks and attempting to carry them around in the air. At 7–8 weeks, we fit them all with light-weight web collars to accustom them to having something around their necks. The collars must be fitted and checked daily to make sure that while they are not overly loose, there is always enough space to easily fit at least one finger under the collar comfortably. At first, the puppies will fuss and scratch at the collars, but as they start to romp and play again, they soon forget they are there.

As we work with them, we playfully and gently tug on the collars while we call "kid, kid, kid," which they know is a call to come to us. Their response to the call overcomes their resistance to the gentle tug and they gradually begin to respond to the tug on the collar itself by moving toward us. We have initiated the beginnings of leash training that the puppies will receive as they grow older, and they have accepted a means of direction by the human.

At age 7 to 8 weeks, we are also beginning to teach them the meaning of some very basic commands. As we play with and socialize them, we place them in a sitting position, telling them "sit" or "sit down." "Sit" works best because it is a simple but very recognizable sound which they will not confuse with "lay down". And when they do and stay there for a second or two, they are praised lavishly and the play continues.

At the same time, using the same method, we begin showing them what "lay down" means. In this case, however, don't lay the puppy on its side or especially not on its back. It will think it is being reprimanded. Place him on his stomach and gently scratch his chest to keep him relaxed and willing to stay in that position for a few seconds. Then, after praising and petting him, allow him to go on with his playing. Each time you see him sitting or laying down on his own, repeat the appropriate words and praise him for being such a good puppy. Through time and cheerful (but not continuous) repetition, this too can become a learned behavior at a very early age, a behavior which does not require "obedience training" to develop.

Of course, the puppy must learn the meaning of "no" and what is acceptable and what is not. When the puppies are so young and are just learning human words, the tone of voice should not be harsh or abrupt.

The word "no" can be said in many ways, but at this point, the puppy does not know what it means yet. He only knows by the tone of your voice. If he grabs a shoe or sock, or something that he is not supposed to have, tell him, "No. Mine," and take it away from him, but not in a manner that he would interpret as playful. If he perceives that you are playing with him, he will persist in trying to get that object. Instead, substitute something that is acceptable for him to chew on, put the other object away and go on about your business so that he knows you were not playing with him. Through gentle repetition, the puppy will gradually learn that "no" means he is doing something that is not acceptable. Once the meaning of the word is learned, a more forceful "no" can be used, particularly for repetitive or aggressive bad behaviors. This may be accompanied by forcing the puppy to its back in a submissive posture and holding him there until he accepts the reprimand.

We do not often use this more forceful form of reprimand, however, as that learned response becomes a problem in leash training later on. The animal becomes fearful and submissive and simply rolls on its back and refuses to get up.

As you will discover, leash training too begins long before the leash ever comes out of the drawer and we have already taken the first step with the young puppies in their light-weight collars.

EARLY SOCIALIZATION—WHAT DOES IT MEAN?

Socialization of Wolf Hybrids is a term bandied about by most breeders and owners of the animals. While there is much disagreement on the methods of socialization, all agree that socialization is a very important component of developing an animal's personality. But what exactly do they mean?

To purists, it means that the puppies are taken from the mother at a very tender age and bottle fed instead of being allowed to nurse from their mothers. They insist that by doing so, the psychological bonding will transfer to the human and not the mother. At this tender age, I think perhaps a better term would be "dependency," rather than bonding.

Many of the purists do not carry the socialization process any further than this, feeling that the nurturing process alone is sufficient to develop a balanced and trusting personality in the developing animal.

Others decry the unnatural separation of the mother and the cubs and claim that when a mother is ready to trust the human handling the pups, she will allow it and that a natural trust should be built with the pups through handling and playing with them from as early an age as possible.

When others speak of a "well socialized animal," they are referring to how well the animal responds to human interaction and handling, how well they react to new and different environments, situations, people and animals, while still others are referring to the fact that they were raised with other animals and interacted well with them.

When you hear people describing an animal as "totally unsocialized," the usual translation of that term will be "unmanageable, fearful or aggressive," and most people will blame the owner or breeder for not socializing them as a pup. This is not always the case, however, as pups from the same litter will react to exactly the same environment and upbringing in different ways. They have individual, distinct personalities, just as do humans.

In reality then, what would be the most dependable method of "socializing" and developing a more balanced, accepting personality in an animal?

While I can understand the concept of the purists that a bond must be developed between the puppies and the human, I cannot agree with the method. From a biological standpoint, the puppy needs the colostrum and the maternal antibodies it obtains from the mother during nursing. It needs that natural protection from disease and the natural nutritional fortification for its developing body. It also needs the warmth and security of the mother and the rest of the litter, as they snuggle together in the den or nest. And it needs the instinctive cleaning and massaging of its body and anal glands to stimulate elimination of the wastes. Most owners know nothing about the need for this process or are reluctant to provide this most necessary care themselves.

Additionally, there is a danger in the nursing formulas provided. While most canine milk replacement formulas nowadays have been corrected to make up for the deficiency, many owners, especially first time breeders, do not know that human baby formulas, or straight milk, or homemade formulas do not contain the ingredients necessary for full canine development. A formula that is deficient in the amino acids afforded by wolf's milk will cause the animal to be blinded by cataracts, as has been documented by Jim Rieder in a five year study to determine the effects of supplemental milk formulas. In that study, it was determined that the milk replacement caused cataract formations during the critical time of lens development. These lens opacities may develop as early as seven days after the pup is placed on the formula. (See Vainisi, et al, 1981, "Nutritional Cataracts in Timber Wolves," Journal of the American Veterinary Medical Association, Vol. 197, No. II).

Many owners do not know how to bottle feed, and often try to cradle the puppy on its back, placing the animal in an unnatural and frightening position, while at the same time forcing the animal to swallow gulps of air as it nurses, if it accepts the bottle at all. These people would do well to watch the bitch nursing her litter to get a better understanding of how it should be done.

From a psychological point of view, I feel that to become a well-adjusted animal, puppies need the comfort and protection of the mother and the company of the rest of the litter and that if the mother is well-socialized and trusting herself, she will at the proper time not only allow, but encourage interaction with the human as she tires of nursing.

To return to human interaction and handling, assuming you have a good relationship with the mother, this should come as a natural phenomenon. It is important that as soon as the mother feels comfortable with him doing so, that the owner begin continuing and frequent daily interaction with the puppies. The puppies do not need to be picked up and forced to accept their handling, they can be touched, scratched or gently massaged in their natural position. The message they should be getting is that the human is not going to hurt or frighten them.

Very shortly after their eyes open and their vision begins to clear, the puppies become very curious about their new world and what is in it, and as they begin to explore and to associate the owner with the always welcome food dishes, if the owner spends time playing gently with them after each feeding, they will quickly respond to this new association. I feel that is very important to allow this acceptance and trust to develop naturally and without being placed in a forced, artificial situation.

But socialization doesn't stop there. It must continue throughout the animal's life. The growing animals need the love, attention and interaction with their owners to reinforce the initial trust that has been developed. They need the companionship of the human and if the owner is required to be gone most of the day (as most people are because of their jobs), the companionship of another canine of comparable size will definitely help ease the isolation the Hybrid will feel if left alone.

A very important element of true socialization is to reduce the natural fear of strange people, strange situations and strange environments. As early as the puppy will accept it, he should be gently and gradually introduced to different people—people who have been instructed in advance not to pick it up or put it into an unnatural position "because he is so cute, they just want to hold him." Likewise, the animal should be taken to different environments where there are new sounds and smells, and while it may be fearful at first, with consistent gentle reassurance of the owner, and if there is no force or insistence that he approach whatever he is fearful of, he will gradually become accepting of these differing environments. While he is still fearful, he should never be forced to endure the presence of, or taken directly to the source of his fear for more than a few minutes, but rather he should be allowed to gradually build his confidence and acceptance of this new and unknown environment. While some this call this a form of socialization, I call it conditioning—conditioning to develop self-confidence and reduce fear.

As stated above, socialization obviously doesn't stop as the animal matures. If an animal is deprived of a healthy relationship with people, it will eventually begin to distrust them. If an animal has no interaction with people or another compatible canine at all, it will develop undesirable behavioral problems.

Socialization is important, and should never be neglected. But as to the definition of the word, you'd better ask the person using it.

"Cidaq." Photo: Joan Fleetman

LEARNING THE LEASH

D.J. Stover

Elementary Education

Leash training, in a sense, can begin when the puppy is seven to eight weeks of age. I say "in a sense" because, as you will remember, it must begin with accustoming the puppy to wearing a collar.

For its first collar, you should choose a very lightweight web collar which has a lot of growing space. It should be checked daily for adjustment and it should be loose enough that you are easily able to slip at least one finger between the collar and the neck. It will take several days to a week before the puppy begins to accept having the collar on without fussing or scratching at it, but as he becomes used to this new "apparel," will soon become oblivious to its presence.

As the puppy grows a little older and is becoming less bumbly in his walking or running, he should also be building more self-confidence and trust of his human companion. Love, patience and understanding of the canine psyche is always important, and especially at this stage of the animal's growth period.

The introduction to the leash should not be a sudden snapping on of this foreign restraint and starting off for a walk. The puppy has had no experience with an additional foreign trapping attached to its collar, and certainly not with its restraints and the new commands that come with it. The puppy should be allowed to get accustomed to it gradually.

Get a cheap, lightweight web leash just for this purpose, as it is likely that it will receive some chewing and other destruction before you have accomplished its first purpose. It should not be a chain leash, as the rattling of the chain may only alarm the puppy and make him resist any contact with it.

Start by carrying it with you, showing it to the puppy but not yet attaching it to the collar. When the puppy is comfortable having this strange, snakelike thing nearby, snap it quietly and gently to the ring on its collar and let him drag it around with him. Guaranteed, when he loses his fear of this new object, he will probably play with it, chew on it, pull on it or dash madly around the yard. Remember, the object of this lesson is not one of restraint, but to get the puppy to feel comfortable and unafraid.

Hook the leash to the collar several times a day and give him plenty of praise and reward as he progresses through each phase of his new training.

As the puppy grows accustomed to having the leash dragging behind him and sometimes getting snagged on things, it is time to teach him that the leash is normally also attached to the human and that it is not a punishment, but a guide to the actions and behaviors that the human wishes him to perform.

Begin by gently snapping the leash to the collar, only this time, keeping a grip on the leash. Let the puppy wander along with you slowly without restraining him or forcefully trying to make him go where you want him to go. He can now be teased into following you by using the old call of "kid, kid, kid" and introducing him at the same time to "Come, Rocky," (or whatever his name may be). He is still a young pup, and if his relationship with you has been a pleasurable one for him, he will normally follow you wherever you go.

Gently try to guide him to staying on your left side, as when we move on to a more advanced form of leash training where this is important, his inclination will be to stay to that side of you.

Do not try to restrain him to staying right by your side however. Let him become accustomed to being attached to you first. This, like all phases of training, should begin with a few minutes at a time, several times a day, and each success must always be rewarded with praise and occasionally, but not always, with a treat. The pup soon begins to realize that even this response to a command or call will win pleasurable recognition from the owner.

If the puppy is suddenly startled or frightened by something, do not pull back forcefully on the leash or force him to approach whatever startled him. If you did, he would probably throw himself on his back, squirming and refuse to get up or respond to you. Instead, go to him and comfort and reassure him in a calm and quiet manner. At this early date, it is more important to build his confidence in himself and his human than to demand obedience.

As you increase the length of your little walks each day, make sure they are not always on the same route, even if you are just walking him within your yard. And, as you are gently guiding him to staying on the left side of you, you can introduce a new word—"heel." As yet, this should not be given in the form of a command, but in a cheery voice saying over and over again, "Heel Rocky," and rewarding him as he responds.

Keep in mind that the time frame of these suggestions is not absolute—that all depends on the puppy's physical development and the speed of his ability to accept each new phase of his training. And remember, at this early stage of his training, we are primarily concerned with developing his trust, self-confidence and understanding of the words his human is saying to him.

When the puppy is responding somewhat to the "come" and "heel" commands on his web collar and leash, it is time to move on to some more sophisticated training and equipment.

Introducing the training collar

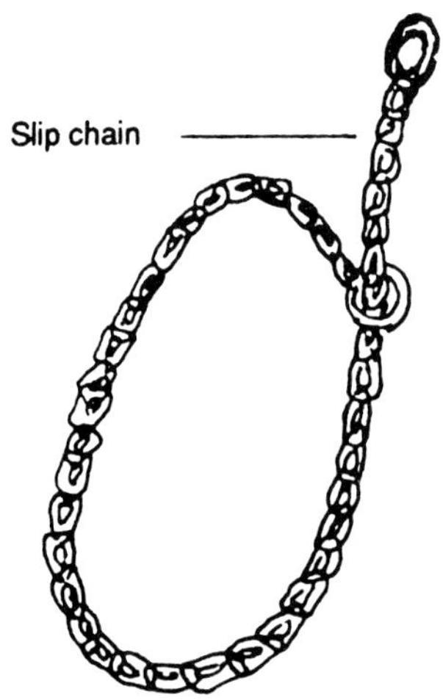

I strongly advocate the use of a "choke chain" collar through the rest of the animal's training. The main reason I do so is because as the animal is learning what is expected of him, he is simultaneously utilizing his sense of hearing to alert him to the movements and commands he can anticipate.

Many people ask me if a choke chain is not cruel. It is not, if it is used correctly and in a manner that the animal can understand. Additionally, the collar should only be put on the dog when it is time for his training period so that he associates the collar with "school time" or later, going out for a walk or run. He knows that it is time to pay attention and respond to the commands if he wants the treats and praise.

Before one can begin to accustom the puppy to the new collar, the human must first learn a little about it. The collar itself should be of a heavy enough weight to restrain an animal of the puppy's size and weight, and yet not heavy enough to frighten him with this new weight around his neck. The collar should be long enough that it slips easily over the dogs head, plus gives three to four inches extra for the "slip" part of the collar. In other words, if your animal's neck when measured very loosely is 18", you will need at least a 22" inch choke chain—and probably an additional inch or so to allow for growth.

There is a right and a wrong way to put the collar on. You have been gradually conditioning your animal to walk or run at your left side for a very good reason—to give you more control of him should that become necessary. It is very important that the collar loop should "slip" towards you, so that when there is no need for restraint, the "slip" part of the chain, that is the end that the leash is attached to, is closest to the left side of your body when he is facing front and away from you.

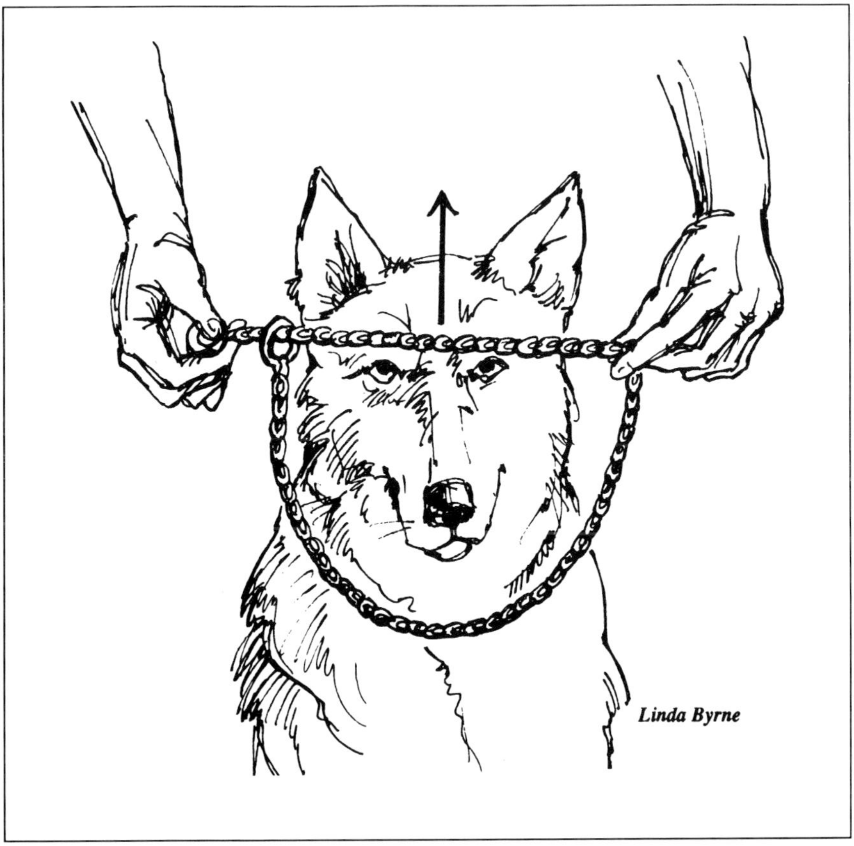

When you are ready to put the collar on him, hold the collar in the manner depicted in the drawing above. Slip the collar on and attach the lead. When there is no tension on the lead, the "slip" part of the collar should easily slide through the O-ring, exerting no pressure on his neck.

Of course, before you actually put the chain collar on the animal, you will once again have to gradually accustom him to this new "gadget." You should let him smell and listen to it before trying to place it on him, as the new weight and the jingling of the chain may at first frighten him. Lay it on the floor or ground for him to sniff and investigate, walk with it dangling from your pocket so that he becomes accustomed to the sound of it. Finally, when he is somewhat comfortable with it, the collar may be placed around his neck in the correct fashion.

Choke chains should never be left on an animal except for training sessions or when the animal is on the lead because the danger of accidental strangulation. If you do leave it on the animal for a short period of time while he is getting used to it, make sure that the two O-rings at either end of the chain are clamped tightly together, and keep a close watch to make sure he doesn't get his paw or some object entangled in the collar.

Continue your walks as before, this time attaching the leash to the "slip" end of the chain collar rather than the web collar. While he may for a time balk at the unfamiliar sound of the chain slipping. You should praise and reassure him as you walk, still conditioning him to the word "heel" as you work and each time he stays to your left and near your side, make sure that the chain is allowed to slip.

Hearing the chain slip as it loosens, gradually conditions him to the message that pressure will be relaxed as soon as he responds correctly to the commands given. But do not apply forceful or sudden pressure - he is still getting used to the sound of the chain and how it works. When he moves too fast or too far from you, the collar tightens; when he is responding well, it loosens.

"Sit" as part of leash training

If you followed the suggestions in the section "Obedience Training Begins At Birth," your animal already knows the command "sit" and responds to it.

"Sit," when on lead, is very important and the training to do so off lead should be continually reinforced to make this part of the leash training easier.

The importance of "sit" when on a leash should be obvious. If you are approaching an intersection or road crossing and must wait for cars, the animal will be much more controlled and relaxed if he is sitting while you wait for an opportunity to cross. If you are talking to someone you meet, you will want Rocky to sit quietly until you are finished; and he will be anticipating the "heel" command when you are ready to move on.

LEARNING TO HEEL, SIT

In your yard or in a safe, familiar place, select an area that is level and at least fifty feet in length. You should use this training area every day until "Rocky" begins to fully understand what you expect of him and to respond accordingly. Here he will learn gradually to stop, go, heel and sit without verbal command. It is in this phase of the training that you must be especially consistent and firm, but at the same time be as gentle as possible.

Attach the leash and take him to the point where you want to begin. Have him sit, and do not move for a minute or so. Then, abruptly give the command "heel" as you simultaneously step off with your left foot. At an even, determined and fairly rapid pace, walk about 30 feet, turn abruptly to the left and stop. Immediately have him sit. When he sits, praise and pet him.

He is obviously going to bump into you when you abruptly turn left and it is helpful to raise the left knee somewhat as you are making your "left face" so that he doesn't knock you down. In time, as you repeat the lesson, he will become more and more watchful of your legs and actions and will begin to anticipate them.

Don't give any verbal commands when you suddenly turn and stop. He must begin to learn to watch your legs and listen to the sound of the chain as it tightens or loosens. The only verbal commands you should give in these lessons are the "heel" command as your start off while he is in the sitting position and "sit" when you have stopped.

Repeat this exercise five to six times in reverse directions and then let him relax for a bit. Later in the day, repeat this lesson, always letting him know what a good, smart dog he is as he begins to respond to your unspoken commands. As he begins to understand and anticipate your actions, lengthen and shorten the distances you go before turning and stopping. You don't want him to associate the same location to the turn and stop each time.

When he seems comfortable and responding to the left turn-stop, it is time to begin turning to the right instead of the left. This will probably involve a little firmer jerk on the leash the first few times, but again, he will begin to anticipate the turn from the tautness of the collar. These turns should also be abrupt so that he once again focuses on your stride, your actions and the tautness or slack of the collar.

Repeat this lesson five or six times and then let him relax. Remember, training sessions should not last more than ten minutes at a time and should be repeated several times a day. As he learns, you will alternate the left and right turns to keep him focused and gradually introduce the "about face" where instead of stopping, you continue walking in the opposite direction.

This particular part of leash training should be repeated and reinforced throughout the animal's life to assure that it remains ingrained.

SIT, STAY AND COME

We are assuming that "Rocky" has mastered the "sit" command and that he will remain in that position at your side when you are working with the turn/stop exercises. He should remain sitting until you step off again and give the command "heel." Now we can apply this training to another situation.

Attach a much longer lead to his collar and walk him to an open spot where you will stop and have him sit. Now tell him "stay" and while you are holding up your open hand, palm toward him, slowly back away a few paces. If he remains in that position, say "come" and lightly tug on the leash. Then give him lavish praise and repeat the exercise. If he does not remain sitting until you call him, do not praise him. Tell him "no, you must stay" and walk him back to the same spot and go through the exercise again.

As he begins to understand and obey your commands, you will gradually increase the amount of distance that you back away from him. When he is really responding well, you may turn your back on him and walk slowly away from him, still keeping a grip on the lead in case he bolts.

This lesson will take a lot of time and patience on both your parts, as he will most certainly want to walk along with you and not be left behind.

When you have both developed confidence in his willingness to accept this part of his training, drop the lead on the ground and walk a little further away. Then walk back to him and praise him lavishly when he has stayed. This exercise will also have to be reinforced periodically throughout his life in order to keep it firmly imprinted.

With this lesson, you have already begun teaching him to "come." You can reinforce the response to this command when you are not in training sessions but are letting him roam rather freely on the long lead. As he explores the area on the length of the lead, call his name and say "come" while gently tugging on the lead. When he does so, give him lots of praise and perhaps a treat. He may even respond to the command without the tug on the lead, which is the goal you desire.

Training is a ongoing experience that begins at birth and continues throughout life. Positive reinforcement of positive responses is essential throughout and although the training sessions need not be continued on a daily basis once Rocky is responding appropriately and consistently, continued occasional reinforcement must be practiced to keep this training imprinted.

OFF-LEAD TRAINING

Your puppy should by now be responding well to the collar and leash, and to the commands you have been working with to sit, heel, come and perhaps doing quite well with the command to stay. When you have taken him to a certain spot, have had him sit, and told him to "stay" and walked some distance away without him attempting to move or follow you before giving the command to come, we have reached a major plateau in his training. Remember, to date all of our training has been done on leash or a long lead, and using the sound of the slip chain collar to enhance the meaning of the commands he has been given. Now we can move on to, for him, a much more confusing part of his training.

While I do not endorse or approve of allowing Wolf Hybrids (or any other dogs) to free-run when they are not in a secure enclosure of one sort or another, and certainly not in areas where other animals, children or adults may be present, certainly in every animal's lifetime, there are unavoidable situations where control off-lead is essential. It may be a situation so simple as the leash not being securely attached to the collar before some irresistible attraction catches his attention; or perhaps begins to dart into the path of a moving vehicle, even in his own fenced yard. And of course, the worst scenario is that he escaped from his own enclosure and is running free, subject to all manner of danger, including that of being shot. It is for these reasons that I feel that off-lead training is especially important for Wolf Hybrids.

But as I said before, this phase of his training is going to be especially confusing to him and will require a lot of time and patience on the part of the owner.

You may feel that he already knows his commands and is obedience trained for those commands. But, is he really responding to your verbal commands, or is he responding to the sound and pressure of the slip collar and the use of the leash? Remember, we have been using these tools extensively in our training.

When "Rocky" has reached the point where the lead no longer has to be employed in getting him to respond reliably to the heel, turn-and-sit exercises, sit, stay and come commands, he is ready to accept the more difficult off-lead responses to the same commands. At this point, it becomes quite helpful to begin using verbal commands as well as his ordinary training collar, web collar and leash. He has learned to anticipate your commands, to a point, through your body actions and the sound of the slip chain, and now he must adapt his entire training to your body language and the sound of your voice only. In other words, we are going to subtract the language of the slip chain and substitute the language of the human voice.

I am sure you are asking yourselves why we didn't use the vocal commands previously. The answer is very simple. If he did not first learn the language and power of the slip chain and leash and did not learn to anticipate your movements through body language, he would not have the required responses imprinted as effectively or reliably.

The vocal commands must always be consistently the same for each command that is given. Too much talk mixed with the commands will only confuse him. A simple "Rocky" to catch his attention and "sit" or "Rocky, heel!" should be sufficient and to him, much clearer than "Now

Rocky, you stay by me and heel." A simple "No!" is much better than "Don't do that!"

So where do we start? Exactly where we did before, only this time, although we put the slip chain collar on him to remind him that this really is a work session, we attach the lead to the light-weight leather or mesh collar instead of the slip chain collar.

Begin the heel and sit training exercises as before. Starting from a sit position, stride off briskly and at the same time tell him "heel!" (Remember that an acceptable "heel" position is no more than two feet from you, and by now, he should be preferring to stay slightly behind you and to the left in order to anticipate whatever sudden moves you may make). At an unexpected point, abruptly make a left turn and stop, telling him "sit." For awhile, he is going to be like a bumbling puppy again, bumping into you, perhaps trying to leave your side, and this is because he no longer can depend on the sound of the slip chain to alert him to your intentions. It may seem to you that you are starting all over again from scratch, but once he begins adjusting to the lack of the announcement by the slip chain and to your new verbal commands, the process will proceed quite quickly.

Again, you must be consistent in your commands and in the way you handle yourself when working with him. This is as much of a challenge and learning situation for you as it is for "Rocky." And don't forget plenty of lavish praise each time he makes a good effort.

Proceed through the same training exercises as were presented in "Learning the Leash," using as little leash pressure as possible, and practice especially the stay-come commands. When "Rocky" is as comfortable and responding as reliably to this collar and lack of pressure on the leash, he is ready to begin working on his ultimate test.

Take "Rocky" to his usual starting point and have him sit. As subtly as possible, remove the leash from the collar, but carry it with you, dangling slightly at your side. Proceed through the same exercises as before, using your vocal commands and trying your best to use the same body language and vocal patterns as before.

Things may (or may not) proceed unexpectedly smoothly for a period of time until he is suddenly distracted by something nearby and realizes that he is no longer under your physical control. At that point, he may break his heel, sit, stay or come command and go happily wandering off to investigate whatever caught his interest.

Don't be discouraged—anticipate that this is going to happen, and while you may scorn and verbally reprimand him, do not punish him. Attach the leash to the renegade once again and repeat the same exercise several times, giving lots of praise for obedience. Then detach the leash and try it again. If he does it correctly this time, really pour on the praise and congratulations, remove the slip chain training collar and take a long, happy break.

Why continue to use the slip-chain training collar?

The slip-chain collar continues to be an important tool throughout the dog's training and reinforcement periods, and throughout life. When the collar is put on, he associates the collar with the fact that this is "school time," a "no nonsense time" and that play time comes afterwards. If the collar is used correctly and the training is done conscientiously and without unnecessary physical force throughout the training, the collar soon becomes a welcome sign that he will get lots of praise and attention with each step of progress that he makes, and in off-lead training, this particular form of psychological conditioning is especially helpful.

Stay-come

The commands to stay or come when off lead will undoubtedly be the most difficult commands to teach with any sort of reliability. It will very likely take the most time and patience in the complete training process.

"Rocky" may be sitting and staying as commanded, when he becomes bored or distracted and nonchalantly breaks the stay command. He should be brought back to the same spot without praise and given the stay command again. Do not expect his obedience for long periods of time, particularly in the early stages of training. Once he has learned what the command means, when he disobeys, particularly off-lead, it is not altogether disobedience that you are competing with. It is more a matter of keeping his complete attention and taking care that the training sessions aren't so long as to bore him. Wolf Hybrids are particularly curious animals and if there is something very distractive to him, perhaps a cat casually peeking around the corner of a tree, I would have to guess that most Wolf

Hybrids would forget about the command and head for the cat. This is not to say that there are not many, many Hybrids which are intensively trained and nearly 100% reliable in carrying out their commands or duties, but one should keep in mind that even highly trained hunting, show or attack dogs do occasionally break command, much to the chagrin of their trainers.

The command to "come" when a dog is off-lead or free-running may seem like it should be "a piece of cake"—after all, it is trained that if it comes, it will get a wonderful treat. Right? Wrong!!

"Rocky" has zeroed in on a gopher hole and is digging furiously. His owner calls for him to come. The most common scenario is that "Rocky" doesn't even hear the command because he is so focused in on what he is doing and the frustrated owner has to go and retrieve him with the collar.

The scenario could be worse—much worse. He could have zeroed in on some small children playing nearby, or on some other animal that he perceived as threatening or encroaching on his territory or space. He could be attacking someone or something.

Repeated reinforcement of the "come" command is essential throughout the dog's life. And it is a command that will probably be the most difficult to enforce reliably.

My experience with Wolf Hybrids has shown me that they, in particular, become so intensely focused on whatever has particularly caught their attention that it is often difficult, if not impossible to break that focus until the animal loses interest himself. This is one reason I am not in favor of taking them off-lead outside of their enclosures and especially not where unexpected and certainly unwanted incidents could occur.

The means of training for the "come" command remains the same as discussed under "Learning The Leash." Again, it will be more time consuming, require a lot of patience, and must be reinforced throughout the animal's life. Treats and rewards are helpful, but more importantly, the animal must have developed a complete confidence, both in your judgment and in your "alpha" status, and must, throughout its life and training, know that it will receive your praise and love for performing the commands that you ask.

Punishment for non-response to the commands is not comprehended by the animal, particularly if he hasn't yet understood what you want him to do. Love and praise are always understood and desired.

“Lone-Wolf’s Moon Shadow.” Photo Jean McCarthy

CHAPTER VI. UNDERSTANDING AND DEALING WITH BEHAVIORAL PROBLEMS

This chapter may seem to be a little disorganized; however, many questions on behavior actually involve more than one behavior. As each situation and each individual animal is rather unique, we are reprinting the entire articles as published in The Wolf Hybrid Times.

"Ahwahnee." Photo: Carol Rush

HEY GUYS–LET'S GO PLAY
A Lesson In Why Puppies Aren't Always Invigorating To An Adult Animal

C.J. Stuart

Tioga was having her first litter. We were so proud as we watched the pups being born—except for one thing. It was 15 degrees below zero outside and she was having one here and one there out in the yard and refusing to care for them or come inside.

Four puppies were born that year and we had to gather them up and bring them into the house because Tioga seemed to have no conception of motherhood, or the slightest notion that they must be kept warm and fed. She didn't even seem to realize that they were hers. Although she seemed to have an ample milk supply, she refused to let them nurse unless we held her down on her side so that the puppies could snuggle up and nurse comfortably. She would tolerate this forced nursing for only so long, and then would make it very plain that she had reached her limit of patience and got out of their reach. In most instances, she would race for the dog door so she could be free of the irritation. Thus began our first experience with bottle feeding.

"What is wrong with this bitch," we asked, as we gently fed and massaged the young infants. "This isn't normal."

Tioga continued to avoid the puppies until they were about five weeks old. They would try to follow her, wanting to nurse and eventually wanting to play with her, but she showed nothing but a slight curiosity toward them. The male was more involved with them than she was. As they grew, he allowed them to climb all over him and played with them almost constantly.

Eventually, when the puppies had been placed in good homes, Tioga returned to her usual personality—the sort of personality one would expect would indicate a good mother—with one exception.

Some months later, my husband brought home a young female puppy that he had acquired from a breeder whose stock was of a different lineage. He liked the temperament of this man's line, and both the physical characteristics and temperament seemed to be inherited quite uniformly. Obviously, this breeder had been working on establishing a reliable breeding line through many generations.

At seven weeks, Mariah was as energetic and playful as she could be. She romped after the adults she was introduced to with glee, and would leap back and forth at them, teasing them to join in her frolic. They all seemed to enjoy this new member of their pack—except Tioga. She would drive Mariah off with snarls and snaps and would growl menacingly whenever Mariah approached.

Tioga's behavior didn't improve with time, and it quickly became obvious that we were going to have two packs instead of one.

We split the half acre compound down the middle in such a way that each pack could still have access to the house when their gate was open, but not have access to each other. Tioga was kept with her longtime mate, and Mariah was with our other two Hybrids.

From that time on, there were few, if any, problems within the pack structure. Tioga was spayed not long after the birth of that first litter. We felt that it was not fair to the puppies she might bear if we did not.

Curiously enough, when we later "babysat" one of her female puppies as an adult, Tioga got along splendidly with her. They ran and played and there was no hint of her former rejection. She just didn't want to be bothered by puppies!

Tioga's behavior suddenly made a great deal of sense to me. Like most couples, my husband and I had often toyed with the idea of having children. But every time we sat down and seriously talked about the idea, we remembered and realized that we honestly didn't enjoy being around children and that in many cases, the behaviors that most seem to find "adorable," simply irritated us. And we could admit openly that we did not want our lifestyle to be tied down by having a family.

I guess to some, that admission may sound pretty selfish, but the point is that not every person or every animal is suited temperamentally to having children or puppies, nor to even being a babysitter or pal. Tioga certainly is not, and I have since encountered other animals with the same attitude. Older animals—fine, but please, mom, keep the puppies away from me!!

MEL'S CORNER

Mel Tanner

Our "Josette" is 72% Timberwolf and Malamute. We really enjoy her. She is 14 weeks old now. My questions are, should we get her a mate? We would like to get another one, but are not really sure what to buy. Such as, is there a particular "wolf breed" that is more even tempered? Would a male and female pack together and maybe turn on myself or my family? If they are outside all day while we are at work, is it okay to keep them inside so much at night? Also, should we get a young male so he would bond to the family even though Josette will be older?

The letter above reflects some questions which I am asked very frequently, particularly by folks considering acquiring a Hybrid for the first time.

By all means, especially if you are going to be gone a good portion of the day, get her a companion. It will save a lot of wear and tear on you, your home and your yards—and her! Like humans, canines need company and "something to do." Lacking this, it is only natural that they would go looking for something to do, particularly at this young age. You might return to find a totally re-landscaped yard or even a new version of "doors" on your house. With or without a companion, you will always be the object of her love and attention, but without trying to sound facetious, there may come a day when you will welcome the companion animal absorbing some of that love and attention—particularly when you're dressed up and she comes racing in through a muddy yard wanting to play. Or if you're not feeling good, or if you just plain want to sleep for a change!

Having a companion will be good for her because of the need for social structure. The standing in their "pack" may change over the years. The likelihood, however, is that since she would be older and has defined the yard and house as her territory, that she would, at least for awhile, be the dominant of the two animals. That could change, but since there would be just the two, and they are of the opposite sex, I suspect that this change would be gradual and rather uneventful.

I would recommend that you do select a young puppy—for two reasons. One, as you suggested, is so that the puppy bonds to you. The second reason is that Josette will accept and bond to a puppy much more readily than if you were to bring a young adult into "her world." Age difference does not matter so much to either of them as that the "new" animal is younger. The younger animal will be the one who will be "taught the ropes," no matter what. Were you to bring an older animal in and expect Josette to instantly accept him, you might be in for a huge surprise. No teenager likes to have someone either their age or older move in and take over their house and their space, no matter how much they might like the person. Dogs react much the same, although a dog's reaction may be much more physical.

So far as a particular sub-species of wolf that might be more even-tempered than another, I believe you are looking at it backwards, particularly since most Hybrids are so many generations away from a pure wolf. For personality, you should concentrate more on what domestic dog blood is involved. Obviously, you would like wolfy looks and conformation, but would like a more docile personality. Before you buy anything, go visit the parents. If you don't like the personality of the parents, don't buy the cubs!

If you aren't planning to breed Josette, you might even consider a plain dog for a companion. I have several pairs in which one animal is just a plain old mutt. I do this mainly because the mutts are so docile and even-going that they help to calm down some of my more high-strung animals. The arrangement works out very well. The mutts have even helped teach the high percentage animals that pulling sled, etc. is *fun* and the Hybrids get very jealous and upset if they can't be included in the team.

So far as keeping the animals in the house, this is one that you will have to "field" as they grow older. Some Hybrids, particularly lower percentage animals, prefer being in the house all the time; others want to be *able* to come in and out as they please, but their preference is to be outdoors most of the time. And still others, particularly those which were never part of the house or family to begin with (adoptees), will refuse to come in at all. I've found that mind go kind of 50–50. They want free access,which we provide via the dog door, but particularly in the winter when it is too warm inside for thier tastes, want to stay outdoors. And now that spring is upon us, with the rapidly fluctuating temperatures (from 25 degrees at night to as high as 65 in the afternoons), they definitely prefer the out-of-doors, as the frozen ground is cooler.

Come summer, they'll spend more time indoors because the house is cooler than outside. But, as I said, this is something that cannot be determined or generalized—each individual animal will develop its own specific preferences which wil be influenced both by how it is raised and its specific environment.

As to the question of whether having two animals would cause them to "pack or tuen on you," this is in large part up to you. It is not a common occurence. In twenty-od years, I have never had such a thing happen. Of cource, I don't go out of my way to do things to aggravate them either. I never play rough with them (which would teach them that this sort of behavior is acceptable), I don't test them by trying to take food or anything else from them; I don't jump into the middle of dog fights or even minor dominance displays between them. When they are behaving very badly, they are disciplined, but in a "wolf way"—grabbed by the neck and forced on theirbacks and yelled at, growled at—sometimes even spit upon, and then totally shunned. It ofen amuses me that when one animal realizes that another has done something "shameful," he or she will jump in and do the diciplining for me.

The point of this long discourse is, don't anticipate problems unless or until they begin to occur. Doing so can cause you to act in a very suspicious manner which may precipitate the very problems you are trying to prevent. Use the same sort ofcommon sense you would in raising a child. Teach by your own behavior, and be consistent. You can teach a dog to be a bully if rough play is allowed or encouraged at one time, but not another

You can also teach them by total disdain that it is not acceptable at any time. Gentle people generally have gentle animals. Gentle, consistent behavior "rubs off" and is absorbed, not necessarily taught and not necessarily inherited.

Still, if when you visit the parents of the puppy you are considering acquiring and don't feel comfortable with their personalities or their looks, don't buy the pup, no matter how cute it is. If you're uncomfortable with the parents, you're probably going to be uncomfortable with the pup.

AGGRESSIVENESS

I have a wonderful, wild, unruly wolf. I live in New York, where laws against wolves are harsh. I need to find a trainer who uses "tender loving care" to train my boy and me in a harmonious relationship. He has taken to dominating me, even biting, growling and barking aggressively. He is one year, three months old. If I could find a Hybrid trainer in any county around me, I'd be forever grateful.

I wish we had received this letter many months ago, before the Hybrid began reaching "puberty." It is absolutely essential that all canines begin receiving their training and that the dominant role of the owner be established early in life. Through lack of early socialization, training and consistent reinforcement, many animals assume the posture you describe—unruly and dominating the owner. The problem is that the proper role relationship was never established when the dog was a puppy, and I would have to guess that you allowed such behavior when the animal was younger, perhaps thinking it would subside.

Many owners erroneously rough-house with the pups, not realizing that while it seems to be play, once such behavior is accepted, it is a difficult pattern to break when the animal becomes too big or too aggressive to handle. What was once acceptable, is no longer acceptable to the owner; but for the Hybrid, it has been a lifelong, acceptable behavior.

You did not give any specific examples of behavioral incidents, so it is difficult to assess what the specific problem may be. Some people unwittingly encourage such behavior by giving in to the animal's every wish and never asserting their own dominance and authority.

Establishing the alpha role of the owner can be learned by observing how mother wolves discipline their cubs. When a a cub becomes out of line, or is perhaps nursing or biting too hard, she will grab it by its neck or throat and force it onto its back in a submissive position, not letting go until the cub stops resisting. With certain puppies this sort of discipline is necessary, especially where establishing thc roles is being resisted.

But dominance doesn't have to, and shouldn't necessarily include rough behavior on the owner's part once the roles are established. Authoritative vocal reprimands, shaming, scorn and subsequent ignoring of the animal (refusing to have anything to do with it for a half hour or so) is very effective, and does not create a challenge situation. Likewise, praise and petting for good behavior is also necessary to reinforce daily the desired relationship with the animal.

It is always useful to the owner to think of a parallel situation with a pre-adolescent child. Physical punishment does little to change a child's behavior. In many cases, it simply reinforces the desire to rebel against what he or she feels to be oppressive authority. With children, one has the ability to discuss their misdeeds with them, to let them understand why certain actions are not acceptable, may be dangerous and express the concern and disappointment of the parent, while allowing for some compromise, provided the compromise is acceptable and agreeable to both.

An animal, however, cannot be reasoned with in this manner. It learns the meaning of a few word commands (such as "No," "Bad," "Good dog," "Come," and so forth), but it must rely on body language and tone of voice to decipher the owner's true meaning of the words.

There are many ways to say "No." It can be said in a firm, dominant fashion, a firm, dominant but shaming fashion, in a casual manner, or in somewhat of a panicky manner. The tone of voice is very important in communicating to an animal what is acceptable and what is not, and what will be tolerated and what will not. Obviously, a high-pitched, loud "No" indicates to an animal that you are frightened and that he is in control. And a "No" that is not consistently followed up by reinforcement, such as shaming or temporary banishment, will not be taken seriously by the animal.

As an example, if a dog steals a piece of food from the table and the owner has reprimanded it with a sharp "No" or "Bad," the lesson is futile if the owner subsequently decides he may as well feed the rest of the steak to the animal, as it is half gone anyway. In this case, "No" or "Bad" simply tells the animal the words have become "part of the game."

As another example, if my own animals try to steal food from the house, though I know they don't understand the words, they do understand the tone of voice, and they are verbally reprimanded severely. They are told in no uncertain terms that what they did was despicable and that it was MINE! They do not receive any more of the food (it will go in the garbage, if necessary), and they are temporarily banished from the house and from my company. Even if I subsequently go out into the yards on some errand, I refuse to even acknowledge the offender.

But you are beyond the early establishment of roles, and at 1-1/2 years, you now face the problem of reversing the roles which both of you have accepted. He is going through "puberty" now, reacting to hormonal changes and exerting his independence. I would strongly recommend having him neutered. Neutering will help to temper his dominance inclinations in time as these hormones come back into balance.

You must, however, immediately begin to correct the role relationship which has developed between the two of you. You must stop conceding to his demands, and through firm, consistent training and behavior, and rewards for positive response from him, establish a relationship where he concedes to your direction.

While it is possible that you need a professional trainer to help you understand what you must do to reverse his behavioral patterns, I definitely do not recommend sending a dog to a trainer or obedience school where the trainer does the training, not you. And I do not recommend training which involves physical discipline. This will only reinforce the animal's perception that aggression is "alright." After all, the owner is being aggressive. Training the owner is 3/4 of obedience training, as it is really the owner who must train the dog, and it is the owner who must be and remain the "alpha"-the teacher, and the provider of the attention and love the animals so desperately need.

Obedience training helps in many ways. It teaches the animal to accept the guidance of the owner and it teaches the owner how not to create challenge situations and how best to manage them with a particular animal, should they occur. Each animal may react differently to any given situation and animals must be consistently "conditioned" to accept and respond to them in the desired fashion. Let me give you another example.

One of my male Wolf Hybrids was extremely fearful of new sights and sounds, especially loud sounds. He would react by trying to escape and occasionally with fear biting. We embarked on a training schedule to gradually eliminate this. Each day, I would walk him down a road where there was a wrecking yard, a very fearful experience for him. We began with stopping to relax some distance away and then returning home. Naturally, he got praise and rewards during each walk. Gradually, we would have our "rest and praise periods" closer to the wrecking yard and after several weeks, I was able to walk him directly into the yard as his fear diminished. He had learned to trust my judgment and knew that I would not force him to endure anything that was frightening or unacceptable to him.

The key to understanding and adapting your animal's learned or natural inclinations lies in understanding their body language, their postures, how they respond to your wishes or commands. If the animal is responding negatively, then you must seek another technique than what you have been using for eliminating undesirable behavior and teaching acceptable behavior.

While finding an acceptable, understanding and gentle trainer may be an option, since most of the training involves education of the owner, I would suggest going to the library and reading various dog training books. Obviously, you do not want the sort that are aimed toward training guard dogs, or which recommend any sort of harsh or physical punishment, but seek out those which are more useful in working with the Wolf Hybrid's temperament.

While the chapter on training in this book is geared for earlier training, the principles are easily adapted to an older animal. The step-by-step procedures will be the same, no matter where you go; however, there may be some steps which your animal has already learned and accepted, and need only be regularly reinforced.

WOLF HYBRIDS AND CHILDREN

I have recently purchased two Hybrids. They are reportedly 75% wolf. Ben is a MacKenzie Valley/Malamute and Sheena is an Arctic Tundra/Malamute mix. They were eleven months and hardly worked with when I got them. They are now 13 months and have come a long way. A friend gave me a bunch of his Wolf Hybrid Times to read and they have been very informative.

I do have one question that maybe someone will know the answer to. I have two daughters, five and six years old. When they're out back playing while the dogs are out, the male will sometimes grab my youngest and pull on her if she's running or going down the slide. We tell him no, but he keeps doing it. Since she doesn't have many clothes on, he sometimes bites her skin, which scares her. He's not putting his whole mouth like a bite—just pulling like he would to a pup. He doesn't do this to my other daughter and they are almost the same height. The female doesn't do anything.

Why is the male doing this and what can I do to get him to stop? My daughter is getting a little braver and turning around and telling him no, but she doesn't always see him coming.

An animal does not look at children the way that humans do. It is more likely to "see" a small animal, even though the stature may seem large in your eyes. The reason for this is in the usually quick spurts of movement of young children, the high pitch of their voices, and the excitability attributable to young children, young animals and small animals.

It is not uncommon to see these responses from any canid, not just wolves or Wolf Hybrids—particularly males—in relation to small children. A child's normal behavior may elicit a sexual response, a prey capture response or the response of an adult animal playing with the child in the manner they would play with one of their own pups. In the latter case, the animal has no way of knowing that this behavior may actually hurt the child. Wolves and Wolf Hybrids are not known to be particularly gentle with their cubs. They do not carry their young in the stereotypical manner, i.e., by the scruff of the neck, but rather pick up the whole body in their mouths, which is not necessarily gentle nor comfortable to the cubs. The child does not know the means of communicating this discomfort or pain in wolf or dog language—simply that it hurts. Unfortunately, the cry of fear and/or pain may initiate the prey instinct in the animal, just as the quick movements of the child may do.

At the risk of offending some of the more gentile, it is also not uncommon to witness a more sexual response from male animals elicited by the children's normal behavior. An animal's sexual behavior is most often stimulated by patterns of movement and partially by sound. If one studies the behavior of the females coming into heat, or already in heat, one may find very direct correlations between their patterns of movement and the innocent movements of children at play. High pitched "giggling" also seems to elicit this behavior.

The very best suggestion would be to not have young children playing in the yards or in direct contact with the animals—especially if they are unsupervised. Young children are very vulnerable and not able to "be in control" and will never be able to be the "alphas." The animals recognize this. The children are, to them, most likely "other subordinate animals." The children may well become the object of a challenge.

It is important to realize that no matter what wonderful qualities your animals may have, they are still animals. But your children are your children, and you can not afford to take the chance of them being reacted to as a lesser or defeatable animal. Far better to keep them separated except under strict supervision until you and the children feel confident that there will be no problem and the children are old enough, capable enough and smart enough to handle the situation.

Please, no matter how confident you may feel about your young children's abilities and/or rapport with your animals, don't take chances. Any large animal with a small child poses a very large risk

No matter what breed of dog is involved, there should be a whole lot of "child education." Even a Chihuahua can inflict a lot of physical and psychological damage and a young child doesn't understand why or what might cause it.

Don't take chances! The children and your animals both need your responsibility.

BITING

Hopefully you can help or advise me. I've owned pure wolves and two Wolf Hybrids for the last twelve years. Although each was unique and had their own personality, a couple of which were difficult, I loved them all and I have never had to seek advice. However, I now have a critter that is beyond anything you can imagine.

He's the lowest percentage I've ever owned and without a doubt, the wildest and most uncontrollable I've ever had or seen. He's a six month old, 110 lbs., 5/8 wolf/Shepherd.

From age three weeks when I brought him home, he's been rough. My arms and ankles were the only things that ever amused him or held his interest and as a result of his razor sharp milk teeth, I was a constant bloody mess. Now, with his permanent teeth, I'm black and blue and my jackets are in shreds. He can't leave me alone. It's always been like this.

No, I don't let him do it or fail to reprimand him. I've tried for six months. I've tried the "No" in a loud alpha voice, I've taken him down and bit his muzzle, I've smacked his butt. I've done everything and he responds as though I'm playing. Nothing affects him. He bites sticks—hackles raised. He runs like he's insane around and around, and each time he passes me, he lunges with his jaws clamping loudly shut. He begins chewing gently on me and then it becomes increasingly harder while he stares in my eyes. He never takes his eyes off the house when I'm in here.

I don't know why, but I'm very attached to this demon so I don't want to get rid of him. I can no longer bring him in the house. At age 3 months, he ate through my walls right into the insulation. Please help.

While I don't doubt that you have tried many things to stop this teenager's rambunctiousness, he obviously requires a different approach in handling.

While you don't say that he is usually tied, the photos you enclosed show him tied to a swivel-type stake. I do not recommend keeping animals tied in any case. It is not at all uncommon for canines of any type that are kept tied to become mean or uncontrollable. In his particular case, he would probably do much better in a large, secure enclosure where he does not feel constrained or tied down and to be kept there with a companion animal.

You do not say whether he is kept with any other canines in a way that they can freely socialize and work their way through "puppydom" together. You are obviously being used as a littermate and if he does not have the opportunity to interact with others of his kind, he does not have the opportunity to turn this activity in a more normal direction, i.e. romping and play-fighting with other canines. If you are the only one he can interact with, he has no other way of working through what would be viewed as "normal puppy behavior" if the behavior was being directed at a littermate.

The biting and chewing on humans must be stopped. There is an old horse trainer trick that, though something of an "old home remedy," just might help. Bake some potatoes till they are good and soft (and good and hot!). Wrap a small towel or cloth around your arm or ankle to protect it and carefully tie the potato onto your limb. Then pull a sweatshirt (or sock, as the case may be) over it and when he goes to bite again, make sure his mouth lands on that potato. It works on horses that nip—they only try it once or twice—it may just work on Wolfy. This would free you of the blame of disciplining him (in his eyes) and would make him rather suspicious of biting humans at the same time.

There is another means of disciplining Wolf Hybrids that I have used over the years. A wolf or Wolf Hybrid is kept in line by the ranking alphas by several means. The obvious, of course, is the continuing physical domination, the pinning down of the betas, the warning "scraps." Less obvious is the total disdain shown by the alpha, while the lesser wolves cringe, whine, beg, even urinate on themselves begging for his favor.

This sort of disdain can be transposed and used by humans as a form of discipline. For example, when an animal is behaving badly or has done some particular thing that he needs to be corrected for, I will verbally and by gesture let the animal know that he has done the very worst, most disgusting thing in the world and then spit right in his face. Then I simply turn and leave and ignore him, no matter how hard he tries to get my attention or forgiveness for the next hour or so. In most cases, the shame and feeling of degradation seems more than they can bear and the lesson seems to be absorbed without rebellion.

Remember, just as with children, there are always those individuals who will not respond to what you consider normal teaching methods and you must try to alter your thinking about how to achieve the responses and behaviors you desire. This often requires some offbeat imagination and creativity. (Some call it reverse psychology). For Wolfy, the "norm" is just not going to apply.

It is entirely possible that as Wolfy matures, a good deal of his hyperactivity and continued puppy behavior will decline; however, it is best to make some positive attempts to speed that decline along before the bad behavior becomes permanent and more serious problems are encountered.

AGGRESSIVE BY AGE THREE?

I'd appreciate more information on Hybrids' development. In the last issue, one article mentioned Hybrids becoming aggressive around age three. More detailed information on this subject would be of interest to me.

This letter was one of several asking questions on the same subject.

One letter mentioned an age of five years. The article to which this letter refers was a highly negative letter to the editor from a former volunteer at Wolf Haven in Tenino, Washington (which was simultaneously sent to many wolf-interest publications).

One must keep in mind first of all, that animals which are seen or heard about at facilities such as Wolf Haven are the "bad eggs," or those which are being displaced because their owners can no longer care for them, deal with them or no longer wish to live with them. What is not mentioned and also ignored in his letter is the fact that the types of animals he refers to represent less than 1% of the Wolf Hybrid population in the United States.

The age of two or three years for behavioral changes to occur in any animal is not a magical number, nor is it even terribly accurate. Worse, the rather misleading statements that all wolves and Wolf Hybrids will exhibit such behavioral changes are as inflammatory as the articles which the "grocery store tabloids" project weekly.

Unfortunately, the press thrives on incidents where children are bitten or little old ladies mauled, and the opportunity to exploit the story even further by the inclusion of the word "wolf" in the article makes such stories even more inviting to them. Factual statistics are not "interesting," such as the data compiled by WHT four years ago on dog bite incidents from four states heavily populated with Wolf Hybrids showing that less than 1% of the incidents involved wolves or Wolf Hybrids. The majority of the incidents involved (you guessed it), your friendly neighborhood mutt of mixed or guessed at heritage. And most of the incidents were triggered by a bitch in heat or the males following along behind.

From where, then, do such dogmatic statements as Wolf Haven's and those of the Humane Society of the United States arise?

As most owners are aware, wolves and high percentage Wolf Hybrids do not reach physical maturity until they are between 18 and 24 months (meaning that they do not attain fertility until that age). Psychologically and physically, they will change and mature somewhat all during their lives, but as in any other species, puberty and a bitch in heat can bring about some rather startling behavioral changes. Just as a young man seems to determinedly test and challenge in order to establish his perceived new status as an adult, so may pubescent animals.

Initial challenges may take many forms—the sudden reversion to "marking of territory" in the house, playing much rougher and seeming more serious in rough housing than usual, an unexplainable growl or snarl, a nip or half-hearted bite, or at worst, a serious stand-off or bite incident.

Generally, if the tentative initial challenges are dealt with correctly, the more serious incidents will not occur. But, what is "correctly?"

Granted, it is not the easiest nor most natural thing in the world to retain one's composure and clear thinking when a wolf is staring intently at you, growling and asserting himself. Yet, this is exactly what one must do. Fear reactions of any sort will only trigger more aggressive behavior. On the other hand, yelling or worse, threatening or striking the animal will do so too. Far better to talk firmly in a normal voice, behave in a normal, non-aggressive and non-fearful manner, to not stare back at the animal and to allow the moment of challenge to pass. Afterwards, heap as much scorn, disgust, shunning and disdain upon it as possible for at least a day and in no event give it any sort of love or recognition during its "grounding."

Wolves and Wolf Hybrids are not dumb, and the loss of your favor or affection for even a day is, for them, the worst punishment you could give. They have not only lost the challenge, but have sunk to a lower rank than they held before. They must learn, just like a child, that no matter what, the human remains the "alpha" and that anything less is simply unacceptable. But one can not teach that lesson by force. It must be by consistent, gentle but firm reinforcement.

This is not meant to imply that all wolves or Wolf Hybrids will challenge their owners in any way. 99% of the animals never exhibit any sort of challenge-type behavior. And, as stated above, age has little to do with it. The time of year does.

As any parent with older children can attest, human teenagers are subject to sudden, unpredictable mood swings and emotionalism. Their body chemistry is changing rapidly and their sometimes seemingly irrational responses to daily life reflect that.

When it is mating season, animals often present the same sort of behavioral modifications. Jealousies and dominance challenges often erupt between animals that were previously very compatible, and sometimes the challenge is directed at the human "alpha." I generally recommend that during the mating season, animals being kept in pack situations be separated into male/female pairs whenever feasible to prevent such situations or that if one is not interested in breeding, that the animals be neutered.

Hunters can easily relate to the aggressiveness of the male elk or moose during rutting season. This is not to imply that the behavioral changes are limited to males. The females are also undergoing dramatic chemical changes in their bodies and it can be reflected in their behavior as well.

The message here is that all wolves or Wolf Hybrids do not present such behavioral changes. In fact, most never do. It is not a breed specific nor species specific characteristic. To say that behavioral changes may occur during breeding season in any species is realistic. It is also realistic to say that such problems can be prevented by proper management.

I am not advocating wholesale spaying or neutering of the wolf and Wolf Hybrid population, although if one is not considering breeding, it should be given serous thought. What I am advocating is logical understanding of why such problems occur and a reasonable approach to prevent confrontations of any sort before they can occur. We are dealing with rather intelligent canines and a little bit of psychology can go a long way.

DEALING WITH DOMINANCE

I am writing to you in hope that some light may be shed on a problem I am having with my Hybrid/Siberian Husky.

"MacKenzie" is a 50% MacKenzie Valley Tundra and Siberian Husky and we bought him at age 2-1/2 weeks and picked him up at five weeks. He was a little devil from the first day he came home, but was very good natured up until about 1-1/2 years ago. He has not come into season (matured) yet, but is showing some signs of this. He loves other dogs and has never been in a fight. He loves all people, except John and myself. He has never been abused or hit, only scolded by our voice.

We understand a lot about wolves and their behavior, but can't quite fit this with our dog. He has bitten both of us many, many times, grabbing with a snap and sometimes breaking skin, but nothing really serious. We both feel he is insecure. He growls all the time at both of us. He doesn't really want to be touched unless he approaches you, and then that doesn't really mean he wants to be petted, just wants to sit next to you on the sofa, or on top of your feet in front of you. He is very good and not usually hooked up on an overhead line unless it's in the evening while the coyotes (during winter) and deer (during summer) are around. He never wanders, and we never have to scold him unless it is absolutely necessary. He is outside most of the days in an enclosure (large one) during the summer and loose during the winter. We have a summer fishing resort and have to keep him there because of the deer and other animals at our elevation of 9,650'.

The confusing thing is that he is so extra nice and sweet when it comes time for his daily long walk, and will even tolerate me giving him a hug without one single growl. This is what's confusing; it's like a game and he thinks he has the upper hand.

He is very good on a leash and he never growls, comes when called–even on long walks, he will eventually come back to where he took off the trail.

We are desperate to find some sort of solution to what boundary we are crossing in his hierarchy so we can make some adjustments, or some solution. I've talked to many breeders, and they all have two solutions; neuter or destroy. I love this animal very much and would never do anything to harm him, but need some sort of help.

I believe that your second to last paragraph states the problem very well. It appears that he does indeed have the upper hand. Like a spoiled child, he has trained you to behave exactly as he would like and to cater to his desires.

One manner of discipline in wolf packs is total disdain, disgust and refusing to have anything to do with the misbehaving animal for awhile. You could utilize this by shaming him harshly and then making him go outside in the enclosure alone and refusing any of his demands for attention, even if you do go into the kennel yourself. If you do this consistently for about a half hour to an hour each time he misbehaves or growls, he will begin to associate this with his bad behavior. Being very social animals, wolves hate to be shunned, shamed, scorned or ignored.

It is clear that he views himself as the "alpha of the pack" and it may be difficult to break that self-image. However, you cannot allow the Mexican stand-off sort of situation to continue either.

If he does not want to be petted, hugged, etc., then don't do it. You might begin teaching him a form of affection that he genuinely enjoys. When you are on your walks and he is more inclined to accept affection, scratch his chest and neck. He may begin to want the same sort of affection at home. Also, when he is being good, let him know that this is desirable behavior, perhaps by giving him a dog biscuit or treat. Always reward good behavior, but you must punish bad behavior in one way or another.

You might also consider getting him a companion animal of comparable size. He would then be able to work out some of his dominance behaviors with this friend, reserving more affectionate behavior for you humans. All in all, I would suggest all three of the above—neutering, consistent punishment for bad behavior and consistent reward for good behavior,and a animal companion. Even "spoiled brat" humans can be responsive to consistent, firm discipline and reward.

COMPANIONSHIP

A month ago we purchased our first Wolf Hybrid—a male, age six weeks. I have to admit, we bought him on the spur of the moment without knowledge or forethought. However, this guy has captivated us in a way impossible for any other animal. We are committed to him and feel it is a privilege and responsibility to be his family.

Our main need at this time is information on raising a single wolf in a family environment not exposed to other wolves. How will this affect his behavior and instincts? The book, The Wolf Hybrid, seems to deal more with people who own multiple wolves. For instance, how will our wolf act during mating season or will he even realize it's that time, not being around other wolves? Does he need exposure to other wolves—like visits back to his original birthplace, or will he be content with just our family? We truly want to do a good job, but we are ignorant and need information.

All animals need companionship; however, this does not necessarily mean that a Wolf Hybrid needs the companionship of another Wolf Hybrid, although the two would understand each other's language best. And, if the Hybrid has the full-time companionship of a human, it does not necessarily need a companion animal.

There are many homes in which one member of the family is nearly always home with the Hybrid and it is a most satisfactory situation. However, I do advise that in homes where the Hybrid is to be alone quite a bit of the time, i.e., during working hours, that a companion animal of comparable size be provided for the Hybrid.

Wolf Hybrids are highly social animals. A Hybrid that is left alone soon becomes a bored and frustrated animal and this is where problems arise. They may become destructive, become "escape artists," or develop personality problems. Hybrids that receive lots of love and attention, on the other hand, are far better adjusted.

With regard to personality changes during mating season, or whether he would even know it was mating season, perhaps a better way of wording it in your case would be "when he comes of age." A Hybrid generally matures at about two years of age. Just as in teenagers, mood swings may become apparent.

It would be my suggestion that since you are apparently not going to breed the animal that you have him neutered now and in this way, avoid any potential problems caused by hormonal changes in the future. Neutering him will in no way affect his personality now and is physically easier on the animal the younger he is. Also, if you decide to obtain another animal at a later time, you will not have to worry about unwanted pregnancies.

You mentioned as an aside in your letter that you had children, the youngest of whom seemed to be eight. Keep in mind that they should be taught not to rough-house with the Hybrid. He may be small now, but you must remember that he will very shortly be a very large animal capable of causing injury just by his size and weight alone.

To go back to your original question, I would not advise taking him back to the breeder's to visit. This will only confuse him and it is unlikely that the other animals will even recognize him as being a former member of their pack. Far better to keep him oriented to your "human pack."

AGGRESSIVENESS TOWARD CAT

I'm writing in hopes that you may be able to help me solve a problem I am having with my nine week old female Hybrid, Savannah. Being 35% Alaskan Timberwolf/Malamute, she is, of course, a beautiful cub, complete with heart-shaped mask. I purchased her from a fine, responsible breeder and Savannah's parents are gentle animals of good temperament.

The problem is that Savannah is very aggressive towards my neutered eleven month old cat, Snow Baby. She grabs him by the neck and when he goes onto his back, she goes for the abdominal area with gusto. Snow Baby loves her and won't use sufficient force to repel her attacks. I keep thinking he'll get fed up, but after four weeks, I'm still waiting. What can I do to nip this in the bud while the cat can still get away from her?

And how should I discipline my cub for negative behavior? Do I pin her to the ground and growl for every negative behavior, no matter what it is? In other words, should all negative behaviors be handled in the same way, despite the fact that some behaviors are more negative than others?

You are right that aggressive behaviors must be stopped before they become "a way of life" and Savannah must understand that you mean business about it. Pinning her to the ground, growling and verbally menacing her while shaking her roughly by the loose skin on her neck until she actually submits is not only appropriate, but should be done every time she threatens Snow Baby in this way. It must be consistent.

You must realize, however, that not every dog can live compatibly with a cat. The quick, scurrying movements of a cat are like those of prey—rabbits, mice, etc., and it is very likely that if the cat itself does not learn some defense maneuvers—like keeping a long distance—there may be disastrous results if they are not kept totally separated.

Not all negative behaviors require the same forceful type of "discipline." Once the animal learns the meaning of the word, a simple but firm and somewhat menacing "No," along with some very obvious shaming will do the trick. As with children, it will depend on the circumstances and the nature of the animal, and as with children, you must use common sense as to what is appropriate.

SIBLING AGGRESSION—SEPARATING THE PACK

I have three wonderful hybrid wolves: Big Sam, 18 months, 105 lbs., 60% Timberwolf/Malamute, and two sisters, Micki and Frosti, 66% Timber/Tundra/Buffalo/Malamute. A very good and loving cross.

Each day is full of different experiences; watching, learning, caring and loving is the name of the game! They are my family. But now the girls have grown up and I have to keep them separated. The girls fight—not just a small disagreement—to hurt or kill! I have tried everything—raining, loving, water, food—nothing works. They both want dominance.

All three have been fixed since six months of age. They are all over one and a half years of age now. Will they out-grow this or will I have to keep them separated forever? Any suggestions?

Unfortunately, once a struggle for dominance between two alpha animals becomes serious, the two animals involved rarely forget their obsession and have to be kept separated for their own protection. An owner's attempts to suppress this battle for supremacy will generally only heighten or encourage it, as added to the need to be alpha, will come the ingredient of jealousy over the owner. If the owner steps in and physically tries to break up the battle, one or both may see this as a sign of betrayal and the bond between human and animal may become badly strained, if not broken.

You have already selected the best means of preventing such dominance altercations—that of separating the two females. I might suggest, however, that you obtain another male companion for the female which is now separated from Big Sam. This will give her companionship as well as help ease the tension and attempts at fence fighting.

Unfortunately, "families that play together do not necessarily stay together" in the wolf world, and in a captive situation, wolves that would normally be driven off in the wild, have nowhere to escape to.

The hardest part for the owner to accept is the feeling of being forced to choose between the two animals. You feel that you are banishing one for the sake of the other. This needn't be the case at all. If you can arrange your containment area so that each pair can receive equal time for your love and attention, all will soon adjust to their new arrangement.

Although I've seen some that do reconcile to a more polite relationship after a few years, the instinct for dominance is not likely to recede in an alpha animal, so it is quite likely that this will indeed be a permanent arrangement. For their own safety and your peace of mind, it would be best to live by their rules in this situation and keep the two females apart.

DOMINANCE

Boy, do we have a problem! And boy, do we need help or advice or something —pretty quick!

The two participants in our active Canis drama are Boomer, our 7 year old German Shepherd (not neutered); and Yukon, our 20 month old Wolf Hybrid male (not neutered), who is 15/16 Timberwolf and 1/16 Malamute.

The other participants include my husband in the role of the alpha male; and Wallydog, a small female generic dog (spayed), who comes to visit, eat, spend a day or two and leave for parts unknown.

The drama

Wallydog comes to visit over the past weekend. She plays with "the boys," eats, sleeps on the couch. The boys are fine. However, when Yukon begins his artful (but silly) adolescent male dominance tricks, such as grabbing Wallydog's entire head in his mouth —or perhaps butting her with his nose, then running away, then coming back for another butt or two—suddenly Boomer starts whining and moaning and strutting, with his tail held high. No violence ensues; Yukon backs away, then returns to play moan-and-groan, bite-my-mouth games with Wallydog. Boomer lies back down on the floor.

My husband decides that he will wrestle with Yukon, which he has done in the past without incident. He and Yukon roll and tumble and wrestle on the livingroom floor. There is an awful lot of verbal communication: moaning and whining and complaining and plain vanilla noise. Boomer comes absolutely unglued. He keeps trying to interfere in the game, but now he's grumbling, and finally, he's making threatening gestures and very deep-throated growls. His face takes on a grimace, teeth showing. My husband keeps telling Boomer to lie down and be quiet, which he does.

Thirty minutes later, the game is over. Yukon gulps down a bowlful of water. My husband heads for the refrigerator to get a beer and sit down in front of the televised football games. Yukon goes into the hallway to lie down on his rug in his place. Everything seems okay.

Suppertime ensues. Boomer gets his chow first (he's the second alpha in the household). Then Wallydog gets her bowl of food, on the floor about ten feet from Boomer's dish. I feed Yukon by hand because he has always been too excitable to sit and eat from a bowl. He gulps a few bites, paces, gulps a few more bites—and generally takes 15 to 20 minutes to eat one bowl. We discovered that when we left dry food out in the large pen with Boomer and Yukon during the daytime, Boomer was dominating both food bowls, and Yukon was getting hungrier and hungrier. We have since maintained the "personalized" feeding schedule in the house, either early in the mornings or late in the afternoons or evenings, because we know that Yukon will get his fair share of food that way. Wallydog does not always come for dinner. She is a transient, but she is a sweetly dispositioned animal who harbors no ill will. Both of "the boys" can tumble her in the snow, bounce on her, drag her all over the yard—nd she continues to come back for more.

This particular evening, after the wrestling game, Boomer began growling from his throat while he was eating his supper. When supper was over, he continued to place himself in his dominance stance between Yukon and Wallydog. We get bones from the local grocery store for all our pets, and there has never been any difficulty with feeding all three animals bones at the same time.

There is no growling, no threatening, no menacing behavior, although Yukon will try to distract Boomer in one way or another so that he can act "wolfy" and steal Boomer's bone. But Boomer didn't seem to care either. He would simply walk over and take Yukon's bone. Tonight, however, we did not dispense bones to anyone.

Yukon is very sweet natured. He accepted Boomer's dominance—and my husband's—up to this one evening. Suddenly, for no reason, he began to growl back at Boomer. We had a lying-down growling communication between the two guys for about ten minutes or so, intermittently stopping when my husband bellowed. No bones. No food. Nothing but evil looks and grimace grins and noise. Everything went well later on. Boomer went to bed in the livingroom. Yukon went to bed on his rug outside our bedroom door. And Wallydog parked herself on the couch.

The next morning was cold—too cold to leave Boomer outside. He spent the day in the house while we went to work. Wallydog wanted to be with Yukon, so she spent the day with Yukon in the pen. When we came home, my husband let Boomer run outside while I fixed the animal food. Then he left Wallydog and Yukon out of the pen.

I was in the house when I heard this hellacious growling and barking and snarling and thudding noise. My husband called me to the door. Yukon had Boomer by the throat; Boomer had inflicted several puncture wounds on Yukon's forearm. My husband, who had gloves and a thick winter coat on, suffered no bites, but he had to practically twist Yukon's collar into a chokehold before Yukon let go of Boomer's throat. Boomer lay still for a moment, then shook himself and got up to walk away.

We examined both of the boys. No one seemed terribly wounded; other than the puncture marks in Yukon's forearm, there were no open wounds. Boomer had lots of Yukon's saliva on his fur, and the saliva appeared slimy and soapy, almost like a lathered shampoo. However, since all three animals had current rabies vaccines, we weren't too concerned about rabies.

We kept Boomer and Yukon separated. We put Yukon in the pen overnight and kept Boomer and Wallydog in the house. No further incidents. Yukon whined and moaned when we let him in to eat; he kept going to the bedroom door, checking out the scents for Boomer. Boomer moaned and groaned from behind the door, but without any passion. Yukon's hackles were up all the time, and they continue to be every time he senses Boomer in the house. He also started spraying the walls, lifting his leg arbitrarily, sometimes as often as two or three times per hour.

Okay, here are my questions:

1. *What's going on? Is Yukon suddenly coming into his own as a "dominant male?"*
2. *Isolated incident?*
3. *Provoked by stray female dog?*
4. *Provoked by husband wrestling with wolf?*

Since the confrontation, we have kept Yukon in the pen. Yesterday, I put Boomer in the pen and went out with Yukon on the lead rope. Yukon and Boomer growled at each other and snarled and made faces and gestures through the chainlink fence. Yukon kept trying to get close to Boomer, but Boomer simply trembled and growled and snarled and kept away from the pen's fence.

My husband thinks we should put muzzles on both animals and pen them up together. I want peace back in our animal kingdom, so that Boomer and Yukon can be the companions they were prior to the confrontation.

Jackie Goodman

At nearly two years of age, Yukon is indeed asserting his claim to dominance and I'm afraid you will have to accept the newly emerging social structure and handle it on Yukon's and Boomer's terms rather than expect that they will conform to your desires. Unless you are willing the expend the emotional and financial burden of constant serious fighting and severe injuries, they should remain separated.

Although some people will try to allow things to follow their natural course, thinking that once one's dominance is established, the fighting will stop, this rarely occurs and after many costly vet bills, most end up providing separate yards and other companion animals of the opposite sex for them instead. You must remember that the animals are not living "in the wild," where the subordinate animal has room to escape and might simply be driven off. They are living in an artificial and closed environment, both with long-standing claims to a specific territory and one female.

Perhaps you could find another "girl friend" for one or the other and keep the pairs separated from each other at all times. One solution would be to section off the yard into three areas—two smaller ones for each pair, and a large running area with access to the house. You could alternately have one pair in the large area where they could come and go from the house as they choose for 8–12 hours and then put them in their smaller pen while the other pair was allowed to be in the "common yard" for another extended period. Naturally, water and protection from the weather should be provided in each area.

This will not change the relationship you have with each animal, but it will certainly prevent the sort of serious confrontations which have happened and will most likely continue if the animals are kept together. Unfortunately, wolves and Wolf Hybrids are somewhat like the proverbial elephant. They don't forget, and once having claimed a dominant posture with another animal, will not relinquish it.

Thanks for putting my dominance questions/letter in WHT. Your response was what I expected, but I'm still going to make one more effort at trying to muzzle the boys. During the first week in March, our Colorado weather dumped 51 inches of snow all over my acreage. Two days later, a second storm brought 24 more inches of snow; and by the end of the week, a third blizzard-type storm, which helped with the drifting, added another 12 inches. I could not leave my house for four days and I got my first (and hopefully last) attack of cabin fever. Naturally, so did all the animals—the ducks, the cats, the dog, the wolfy kid, and our newest addition, a 50% Arctic Tundra/ Malamute puppy who was only five weeks old.

I am going to try to muzzle both of the big boys while there is still snow on the ground, which may (or may not) act as a buffer so that they do not injure themselves. Both Boomer and Yukon have calmed down considerably since the storm and their enforced house cohabitation with me during the four day hiatus of the storm. If nothing works, I have not lost more than the cost of two sturdy muzzles. My veterinarian has agreed to give me a vial of tranquilizer tablets to be given to each animal prior to the experiment. He will be on hand to help me with their reintroduction. I do not want to hurt the boys, but only to try to regulate the dominance factor. Since I will not be able to put up a second run until late July or August, I am concerned that the summer will be miserable for the boys. And that's not fair to them.

Jackie Goodman

While I appreciate your desire to have Boomer and Yukon together once again and be able to "regulate the dominance problem," I feel quite strongly that this is not the answer.

In the first place, they may injure themselves quite severely trying to remove this new restraint. They will most likely view it as some severe punishment and the result may be even greater behavioral problems for both you and the animals.

Muzzles are used as restraints on a very short-term basis *when necessary*—normally if veterinary treatment is required and there is a likelihood that the animal will bite, or if such an animal is being taken around other people where a bite incident might occur. For longer periods, a muzzle is not a good idea.

The animal must be able to eat, drink water, pick up sticks, bones, whatnot. A muzzle is somewhat akin to wiring the dog's jaw shut.

It would be much less stressful to all concerned if you could build or obtain a temporary pen so that each could have time to be with you for long periods of time—but separately. Portable chainlink pens of up to 10' x 12' x 6' high or greater (including a gate) can be purchased from places such as Sears or local lumber or hardware yards. These pens can easily be assembled, disassembled or moved by one person in less than 20 minutes. They can be securely anchored to the ground while in use in the same manner as a tent or trailer is staked. I use such a pen when I have puppies and disassemble and store it in the shed for the rest of the year. Storage space takes up less than three feet for the entire pen.

FEAR OF NEW OWNERS

We have owned our first Wolf Hybrid, Keaton, for only two weeks. He is 88% Alaskan Tundra and 12% Malamute.

I believe Keaton has been beaten. His will is broken. He loves and trusts me. My husband is feeling very hurt because Keaton is so afraid of him.

Keaton is a little over two years old. I keep telling my husband it will take time for him to come around because he has to learn to trust him, (kind of like a woman who has lived with an abusive husband). Time, usually, is the best medicine, but it hurts to see such a beautiful, huge and sweet animal so timid around men. He loves me and the children.

Many people adopting grown wolves or Wolf Hybrids have this problem. Keaton's behavior does not necessarily indicate that he has been beaten, although having been treated roughly in the past by a man is a possibility.

You do not indicate whether Keaton was kept strictly in a kennel, allowed to share the house, kept on a chain, or whether there were other family members involved in Keaton's past. Without these factors, I can only give you possible explanations for his behavior.

No matter what the containment pattern for this animal has been in the past, first and foremost, he is going to have identified with and bonded to the person that cared for him and to his previous surroundings. This may or may not have been a happy situation for the animal; nevertheless, the attachment, if only through need, is there and he has suddenly been removed from whatever security he had (i.e., food, water, perhaps an occasional petting in a setting he was accustomed to).

The removal from an "accustomed setting," no matter how good or how bad, is very traumatic. Removal from the person they have bonded with is even more traumatic, no matter what the conditions of the bonding were based upon.

I have occasionally adopted grown animals, with various degrees of success in developing a "kinship" relationship. It is very difficult, because they have bonded quite some time ago to another individual, and in their eyes, that individual let them down, no matter how they were treated by him or her. No matter how good your intentions, they are suddenly orphans.

It is, of course, much harder if the animal has been moved from owner to owner, where no stability can develop. It helps new owners of mature animals to think of them as adolescent human children who have been bounced from home to a foster home, and perhaps another foster home, etc. The lack of trust of new families and new situations will take a long time to develop—certainly longer than two weeks.

As to why Keaton may accept a female sooner than a male human, that question is easier to answer and is applicable no matter what circumstances he had previously lived under. A woman's voice and temperament are generally more gentle, quieter, and her manner of dealing with children, animals, etc., is generally more gentle sounding than that of a man. A man's voice, no matter how gentle, is deeper and sounds more like a growl to an animal (just because of the deepness of tone), and for this reason, is threatening to a strange animal.

A man is also more prone to look directly into the eyes of a person (or wolf) he is talking to, but to a wolf, staring into the eyes is a direct challenge. A skittish Wolf Hybrid (or wolf) will most often finally make the attempt to be friends from behind, where you cannot look into its eyes.

I have spent many hours in small pens (10' x 20') just sitting with new adult wolves. I don't look at them and they try hard not to look at me. I sing, I talk to myself, (or to them, but they don't really understand that for awhile). They don't like or trust my voice or presence. Usually, I will sit out there near their food dishes—but no so close that they can't come up and sneak some food when they think I won't notice. Sooner or later, they do come near me, but always when I don't look directly at them. Later, when they have approached very closely several times, I turn them out into the bigger yards and they usually start following me around, though always from behind. I don't push it and rarely turn around and look at them, even when I feel nudges on my calves from a nose that seems to want attention.

Sooner or later, after regularly sitting in the big yards with them, the animals will come up behind me when I am sitting in the dirt and stay by me (still, so long as I don't turn around abruptly and look at them). They will follow and nudge me when I'm walking and I can pet them behind my back. One day, unexpectedly, they will "sneak a kiss" while I am sitting. But the crowning glory is the day that they finally overcome the distrust and I can actually look them in the face and give them the affection one-on-one they have been trying to muster up the courage to obtain.

It does take time—lots of it. Some have taken more than two to three months to accept me. But think of it this way, how much longer would it take to get a 12 year old foster child to accept and trust you?

As with children, if you could have nurtured and had this animal from early "childhood" to maturity, these problems would probably not exist. If you have the courage and ability to nurture and bond with an adult wolf or Wolf Hybrid, you will have accomplished not only a feat of patience and love, but have probably experienced the most enduring lesson about love and understanding that God has given humans the opportunity to learn.

PROTECTIVENESS

Are bones bad for Hybrids' teeth? Can they be easily worn down or is it natural for them to eat bones? Also, Kodiak has begun displaying aggressiveness (growling) toward anyone who approaches my boyfriend and I. Friends say that if we neuter him, the "guardian" instinct will subside. Does neutering really deter the natural protective instinct of the Hybrids?

Raw meat bones will not harm the teeth unless the teeth are already soft to begin with. Actually, raw bones help clean the teeth and the chewing helps to strengthen and keep the gums healthy. You do not mention what kinds of bones you are talking about, but probably the best bones to use are the knuckle bones from red meat animals. These will last the longest and do not splinter and usually do not crack or break like other bones.

Poultry bones are not recommended because they not only splinter and fragment easily, but are often swallowed whole by large canines. Poultry bones are notorious for either becoming wedged in the mouth or in the intestine, the latter causing severe problems requiring surgical procedures to remove them and often severe internal bleeding. Obviously, it is better to avoid feeding them poultry bones.

The question about neutering to deter protectiveness is interesting in that it does not describe aggressiveness per se, but protection of his owners. There is a distinct difference.

While spaying or neutering may diminish true aggressive behavior, it is doubtful that the need to "protect his family" will subside. Some animals "protect" their family by getting between their humans and other strangers and pushing them away from the strangers. Others give much stronger warnings. Occasionally, when particularly suspicious, they may even attack. Far better to keep the animal in an area where it cannot come into contact with "outsiders" than to take a chance that it may misinterpret some movement that a person may make as threatening and respond accordingly.

It may be possible to teach Kodiak that certain people are acceptable and that he will eventually accept them. This should be done with care, keeping him on a short leash while gradually over a period of time shortening the distance between you and the person he perceives as threatening to you and do this on many different occasions. Talk to the person in a normal voice, while continuing to reassure the animal that this is a "friend," so that he will come to know the word. It will take some time to teach him the meaning of the word, but with patience and lots of reassurance and petting from you, it may pay off in the end. You should also gently but firmly tell him "No" when he growls, while still petting and assuring him that this "friend" is alright.

Begin with only one friend until he accepts them as such. If he does not, then you should be prepared to simply avoid the situation altogether by not allowing him to be with you when others are present. If he does accept the training, the next step would be to gradually introduce him to a new "friend" in the same manner.

You must keep in mind, however, that no matter how accepting of other people he may become, if they make moves that he perceives as threatening to you or your boyfriend (or perhaps his territory or himself), you may encounter the same behavior and should be on the alert for this. And, of course, he should be made aware of and rewarded for his good behavior at each stage of the process.

Miracles do not happen overnight and you should expect that this process of modifying his basic instinct will take time, and that it will not necessarily apply to everyone you tell him is a "friend." Even if he begins to accept a number of people, there will always be a personality type or visual image that may illicit the same response, just as it does in humans.

Fear or protective responses can be reduced through patient conditioning in many instances, but again I must emphasize that this will not happen spontaneously and will not necessarily apply in every new situation. If the conditioning through gradual exposure doesn't work, keep the animal confined when others are in "his" area.

BEHAVIORAL CHANGES DURING MATING SEASON

I am a proud, loving mother of an over-pampered 2-1/2 year old male Hybrid (62.5% Arctos, 32.5% Giant Malamute) and have been a grateful subscriber to your magazine since our Kodiak came into our home at five weeks of age.

During the course of these past 1-1/2 years, I have read many stories of yards looking like lunar landscapes, personal home furnishings being destroyed or at least found half buried in the yard, etc., and consider myself fortunate that my little ray of sunshine really hasn't proven to be too destructive. Oh sure, we've had several remote controls "self destruct" while at the grocery store, as well as a few pillows disemboweled," but never anything that I couldn't attribute to my own neglect of leaving the house without wolf-proofing it first!

When our Kodiak reached a year and a half, we came to a conclusion that his continual escape attempts were primarily hormonal in nature. When we fenced him in, he figured out how to use the gate latch, or at least dig under. When we tried a tie down cable, he realized immediately how to dig up the stake which anchored him. We determined he was desperately lonely, so we bought a purebred Malamute to satisfy his urge to wander. It worked like a charm!

Kodiak is now 2-1/2 and NaNooki is almost one, and she is ready to go into heat at any time. Kodiak has become completely satisfied living with his "pack" now, and may run outdoors freely without our concern that he will run off because NaNooki is trained not to leave the property.

Our concern is this: Now that NaNooki's scent makes her so attractive to Kodiak, Momma may love on them, but whenever Papa loves on NaNooki or Kodiak (this is during the time when she is close to season), Kodiak bristles and growls harshly. He does wag his tail to some degree, but if it were anyone other than Papa, I would fear he would be in attack mode. I realize he knows better than to snap at Papa, for Papa established dominance a long time ago and Kodiak has always respected that.

I'm writing to see if this is common behavior or possibly acquire an explanation or insight.

At 2-1/2 years of age, Kodiak is simply "doing what comes naturally," and this behavior will reoccur every time NaNookie comes into season. Just as the female's hormonal balance is disrupted during estrus, so is the male's and like in people, when the hormonal balance is disrupted, personality changes occur. You might compare it to the human premenstrual symptoms in the female or the mood swings during puberty or menopause.

NaNooki is Kodiak's mate and he doesn't want any competition for her attention, and at this time, does not want to be handled or distracted himself. Right now, his attention is focused on one thing—the female in heat.

Don't force the issue; don't lavish affection on them at this time if he doesn't want it. This will pass once NaNooki is out of season.

I would suggest one thing, however. If you don't plan to breed the animals, have at least the female spayed now - even better to have the male neutered too. The spaying will obviously eliminate the heat cycle, but by also neutering Kodiak, you will pretty much eliminate the mood swings and season behavioral changes common to an unneutered male. Neutering will not instantly change the hormonal balance, but gradually the physical urges wane and behavior returns to normal.

In the meantime, be patient, consistent and firm, but do not force your attention or demands on either animal and perhaps provoke a confrontation. Without appearing to be weakening in your role, respect his need for aloofness at this time.

Kodiak's growling and bristling is a distinct warning that you are encroaching on his "space," and in this instance, you must respect it. Once a confrontation does develop, many owners sense a barrier of distrust between themselves and their animals, and it is best to avoid it altogether.

Again, if handled correctly, this situation will pass in a few weeks, so give him some time.

“Columbo Chief.” Photo: Terry Hay

CHAPTER VII. MEDICAL PROBLEMS –FROM AN OWNER'S POINT OF VIEW

This chapter deals primarily with dysplasia and the different types of surgery which can now be performed to correct it. Hip dysplasia appears in several forms, some of which will be discussed, as well as causes of this degenerative disease.

While corrective surgery is expensive, it is often the only viable option for responsible owners who are very attached to their animals. The suggestions given for demanding proof that the parents of the puppies are dysplasia-free should not be taken lightly. The breeders must take the responsibility of being absolutely certain that they are not perpetuating this tragic fault in the puppies which they produce.

TUNDRA—AN OWNER'S GRIEF

Carla A. Childs

Tundra came to us in April. He was a beautiful big baby of 5-1/2 weeks old. We had the usual problems of allowing him lots of privacy and time to get to know his surroundings and new pack.

It took a short time, but he was part of us and very much accepted by Timber, our other Hybrid. Everything was going fine, except I noticed he walked a little funny. He was large for his age, seventeen pounds, and I thought he was just a bit clumsy.

Two weeks later, I mentioned this to my vet and asked about x-rays. He felt it was unnecessary at the time, so I called where I had purchased him and asked if there had been any history of dysplasia and if the animals had been x-rayed. The answers were no and no.

I was already unhappy with the vet, as his attitude toward our other Hybrid, Timber, had always been "a dog is a dog," so I began the search for another doctor. It did not take me long to find Dr. Gayle Roberts. She was wonderful. She was interested in my animals from the beginning. The winning question was when I asked her if she felt my animals were classified as a dog. Her answer was "No, a wolf is a wolf and a dog is a dog. You have hybrids."

At that point, I explained my concern for his hips. She checked him, took him out to walk and found that he did indeed not want to extend his right leg as far as his left. Although it was far to early to usually make a diagnosis, she did feel x-rays were warranted, as Tundra was new to me and I did not want to get attached to him if there was a problem.

The x-rays indicated definite signs of a serious problem. She gave me a letter to send to my breeder and I immediately placed a call requesting that she take the pup back. She did not want him, of course, and we talked off and on about keeping him for pet quality and perhaps working out a 50-50 deal for payment of surgery costs, if needed in the future. Dr. Roberts put me in touch with Dr. Hoeffle at ISU who took a look at his x-rays and agreed we might be able to work out something on surgery.

From that point on, Bob and I decided we had the support of people and we would keep Tundra. I was already attached, but I did not realize how much I was going to grow to love him and how painful this experience was going to be for us all. We were in the process of moving, so things were really hectic. We followed up with another set of x-rays before leaving and everything seemed to be okay. Not better, but not getting any worse.

When we arrived at our new location, I again was blessed to find a doctor for my Hybrids that is exceptional. Dr. Fore has become a trusted doctor and friend of my Hybrids. We explained the problem to him and he had Tundra's x-rays sent from California. He concurred with Dr. Roberts and Dr. Hoeffle that there was a problem, but that since his hips were not bothering him, we would just keep an eye on him. He took x-rays a couple of months later. Tundra was ten months then and almost one hundred pounds and he felt they looked worse. We all kept our fingers crossed.

Then the fatal day. Tundra came upstairs to my room and it was obvious that his hip was out. We took him to Dr. Fore and had to leave him. Dr. Fore put the hip back in place but held little hope for avoiding surgery. He kept him confined in the hospital under observation for four days to see if it would improve, but it did not. Surgery was decided on after much turmoil and emotional upset. We had no choice. Tundra was one year old.

Dr. Hoeffle was on vacation and I could not even consider the best surgery, due to the cost—about $1,000 per hip. We had no choice but to go with the one surgery on his right hip. It went well. Tundra was operated on Friday and released to come home on Monday, as he was quite depressed and it showed. He was home after eight days in the hospital. He was in pain, had lost 14 pounds, would not eat.

My whole life was turned upside down. Timber was aggressive with Tundra and I had to keep them separated. I slept downstairs in a sleeping bag, as Tundra could not climb the stairs.

I tried to get some weight on him by feeding him some good stuff with his dog food, like cottage cheese, chicken parts, liver, etc. That plan backfired as he began refusing everything but the goodies, which frustrated me and him also.

His stitches were taken out and two days later, the skin over his hip split open at the top. Another trip to the doctor to have it restitched. He seemed to be using his leg a little when all of a sudden we were back at square one. He had bumped it or something and was walking like he had just come home after surgery.

At that point, I sat down and cried. I found I was angry at him because after four weeks, he was still not eating right, things were not back to normal and I worried that if he didn't start putting weight on that leg, his other hip might soon give out and none of us could go through this again!

I would not put him through the upheaval this has brought to his life, or the pain and hurt it has caused all of us to see him suffer so. And they do suffer—not just the pain of the hip, but the emotional stress and confusion of what is happening when their world is turned upside down.

I am here to say that dysplasia does exist in Hybrids and in some cases is very severe. It is very unusual for a diagnosis to be made at so early an age. This does not usually show up until the animals mature.

If you are contemplating buying a Hybrid, please be sure that you have documented proof of x-rays on both the parents and the grandparents of the cub you are contemplating buying. Do not just ask if there has ever been a problem. If you cannot get documented proof of clear hips, don't buy from that breeder. Also, get a contract that states just what responsibility the breeder is willing to take if by some chance you get a cub with genetic problems. Do not take oral promises because they do not hold up.

In my case, when the problem arose, everyone was surprised. But that does not solve the problem. I am now in debt on Tundra a good $1,000 at least, and even though the financial part of the care of Tundra was discussed with my breeder a year ago, no help is available. She is offering a new cub, but I won't take the risk of buying out of the same parents, or even a different female with the same male. I would be asking for trouble. I have asked for my purchase money back and have been told she doesn't have it.

The point I am making is that if you do not take care to cover the bases, you will most likely be left with a broken heart and a very hefty doctor bill, and no help whatsoever.

I can't find a way to tell you what it is like to try to console a seven year old girl who is in tears and afraid that Tundra may not make it. To see her try to feed him by hand, to pet him, to love him. To describe the look in his eyes as he tries to do things and then whine from the pain. To see him relinquish his right as alpha, which he earned.

If you have a Hybrid and have not had him x-rayed, please do so. I had time to prepare for this. I knew it was probably coming and it still knocked me off my feet. I didn't expect it so soon and had hope it would not happen at all. You are never prepared for something so devastating, but it is better to know if you have a potential problem.

I hope Tundra lives. I can't see life without him.

RUNTSTER—SURGICAL HELP FOR A DYSPLASIA VICTIM

Nancy Burger

"Tundra—An Owner's Grief," an article that appeared in the October issue of *The Wolf Hybrid Times* was one story that has had a special impact on me. At the time I read the article, I was trying to make a decision as to whether or not surgery was going to be performed on one of our Hybrids, a four month old female. I just didn't want our puppy to go through what Tundra had been through.

Runtster was the first born in her litter and until number two came along, we had no idea that she was undersize—the runt of the litter. It was several hours later that we noticed her front feet bent to the sides like flippers and we began to feel she might not make it. The vet said that it was a 50/50 shot that the feet would straighten out, but there would be no way of telling at that point if there were any other internal abnormalities. We decided to see what would happen.

A runt, but not abandoned

The first week was quite stressful because the mother of the litter would not go anywhere without taking the Runt along. We thought at first it was an attempt to abandon her, but found out in a short time that she had no intention of giving up on Runt. I would wake up in the morning to find Kiska (the mother) and Runt in bed with me. Runtster was allowed to nurse whenever she wanted to, even though the others were put on a schedule by Kiska.

After the first week, Runtster was left with the other pups and by three weeks, her feet had straightened out.

She always had a different gait than the other pups and did not seem as sturdy in the rear. We hoped that it would get better in time and that it was only due to the fact that she had to work her back legs more those first few weeks to compensate for the flipper front feet.

At about four months old, she was experiencing some lameness and x-rays revealed bad hips. She had good and bad times, but at six months old, a decision had to be made because she was in obvious pain most of the time. With some hesitation, we decided to give surgery a try.

A femoral head ostectomy was performed on her right side. At the time of the surgery, she was 66 pounds and just a little over six months old. Dr. Heim was pleased with how the surgery went and we brought Runtster home the next day. The size of the incision along with how small the leg looked shaved was enough to make me wonder if we had done the right thing.

My questions were answered within the first few minutes of the drive home. Runtster yammered and talked all the way home. She was not happy about being left at the vet's and let me know she was glad to be getting out of there. When she arrived home, she greeted the rest of the pack through the fence and proceeded to play with her new toys. She was so energetic that I called Dr. Helm to see if she should be allowed to be this active. She wasn't actually using the right leg, but that didn't seem to slow her down. She ate a full dinner, played some more and finally settled down to sleep.

It has been ten days since the surgery and her stitches were removed today. Runtster is ahead of schedule on her recovery. She is doing in ten days what is expected at three weeks.

She is happy, playful and loves to be the center of attention. We were very careful through all of this never to let her know we thought anything was wrong with her and her attitude was always much better than we expected.

Major surgery, but worth it

I know there must be others facing the decision of what to do about their Hybrids with hip dysplasia. Whether it was caused by genetics, over-nutrition or a fluke of nature, the outcome is the same and it is not an easy thing to go through.

We were fortunate to have had such a wonderful recovery so far with Runtster. She will never have a perfect walk, but once she is completely healed, she will not have the pain like before the surgery.

I think it's important to realize what a major operation this is, but that it can be an end of pain and a very successful procedure in many cases.

KARHU—A SUCCESS STORY

Janet Lux

After reading Carla Childs' article, "Tundra—An Owner's Grief," in the October issue, I felt the need to write my own story. It has a happy ending due to caring veterinarians and our commitment to our Wolf Hybrids.

My husband and I live in the suburbs of Denver, Colorado—not exactly where you would expect to find a three-member wolf pack. We have a large yard with a 7-1/2 foot high fence, trees for shade and a patio with a dog house for protection from the weather. Every weekend is spent in the mountains; hiking, back country skiing or fishing, and our wolves are always with us. Our pack consists of Heidi, age nine, 50% wolf and 50% German Shepherd; Sultan, a 3-1/2 year old registered Wolf Hybrid—5/8 Timberwolf and 3/8 Siberian Husky; and Karhu (pronounced kah-roo)–7/8 Alaskan Timberwolf and 1/8 Malamute, age 11 months.

At the tender age of seven months and weighing 85 pounds, Karhu was operated on for severe hip dysplasia in his right hip. Now, four months later, he is completely healed, runs and plays with his wolfmates like a healthy 11 month old 100 lb. puppy should. There is no pain or stiffness, even after strenuous play.

Hiking season is short in the Rockies, usually July, August and maybe a week or two in September. We usually hike the rough terrain wilderness areas, to find the solitude my husband, Fred, I and our "kids" thrive on. Karhu was only five months old when hiking season arrived, but very large for his age. He naturally enjoyed the outings and we felt he could handle the long 5 to 8 mile hikes we went on. After a few of these hikes, he started having pain and stiffness in his hips.

We cut down the lengths of the hikes, thinking too much, too soon. But the week long rests in-between hikes didn't help. He got worse with each hike. He would crawl over logs rather than jumping over them. He never did get up on the furniture like a normal pup. It was as if his hips simply weren't strong enough. When getting up, he would arch his back for the first few steps.

It was *deja vu* for us. We had seen these very same symptoms before. Less than a year before, we had to put our 12 year old German Shepherd to sleep because the pain from his hip dysplasia was so intense. When he was less than a year old, we suspected he had the disease, but at that time there was no cure. All surgery was strictly experimental at best and very expensive. And here is our newest baby, a mere seven months old, looking and walking just like that 12 year old dog. It had become frighteningly obvious that our highest percentage wolf yet had hip dysplasia. We reluctantly took him to the vet, expecting the worst. X-rays were taken and the vet confirmed hip dysplasia.

Dr. Friesen referred us to Dr. Rouse, a surgeon. He looked at Karhu's x-rays. Our pup had hip dysplasia, only 30% contact in his right hip, but there was no deformity in the socket or the head of the femur. And he was the right age, 7–8 months old. The bones were still growing. All those factors made Karhu an excellent candidate for a surgical procedure called a triple pelvic osteotomy. The success rate was very high.

The operation involves cutting through the pelvic bone on either side of the socket. This piece of bone is then turned so that the socket fits over the head of the femur (its proper position). Steel plates and pins are used to hold the bones in place, so the bones grow back together in the right position. The plates and pins are permanent, never to be replaced or removed.

Dr. Rouse then advised my husband that it would be expensive, around $650.00 at that time. Would that be a problem? My husband's instant answer was "no." We had watched helplessly as Max suffered in his later years. Today there is a surgery that can prevent all that suffering. Yes, we wanted the surgery, no matter the cost. Our wolves are our children, as precious to us as a human child would be. The thought of letting a dollar sign get in the way was ludicrous.

Karhu was scheduled for surgery the following Tuesday, seven days away. One last weekend camping trip before his three hour operation. After setting up camp, we went on a short walk. After a quarter mile, Karhu was limping and hurting again. The aspirin tablet helped, but we knew our puppy was ready for surgery. He could not go on like this.

At 6 p.m. Tuesday evening we received a call from Dr. Rouse. He said the operation went well and Karhu was awake. We could come get him on Thursday evening.

The next two days were torture for us and the other two wolves. Sultan, the male who had adopted Karhu as his own puppy, in typical wolf behavior refused to eat and kept searching for him. This alpha male was actually angry with us. After a day of this, we sat down with Sultan, *a la* human child, and explained that Karhu was coming back. We got out Karhu's leash and kept repeating his name. It seemed to work. He relaxed and ate some food that night. How human these wolves are. Anyone not familiar with these incredible, intelligent animals would have thought we were crazy.

Thursday afternoon we went to get our Karhu. Sultan could sense it. When he saw Karhu's leash in my hand he came unglued, doing everything but backflips. He knew.

We waited forever, it seemed, for them to bring out Karhu. He was limping badly, sporting a 6 inch, 13 stitch wound on his right hip. His ears were down, eyes half closed from the pain. All the hair had been shaved from his right hip and leg, and the front of his other legs. When he saw us, he wagged his drooping tail once, wobbled over to us and let out a string of yelps when he laid down. We were sent home without antibiotics or painkillers and instructed not to let him walk too much for the first week and limit him to leash walking for two additional weeks.

We brought him home. Fred lifted him out of the truck. Every movement brought new cries of pain, but Karhu was glad to be home.

The reunion was grand. Sultan and Heidi sniffed him from nose to tail and especially checking out his wound. All Karhu could do was wag his tail and lower his head in wolf greetings, but he really wanted to be left alone. Once over the initial excitement, Sultan and Heidi understood his snapping at them and kept a close eye on him—from a distance.

Because of the extreme pain, Karhu got to his feet only when it was absolutely necessary. For the first few days he would urinate where he lay. It wasn't worth the pain. If he was laying inside the house he would whimper, trying to get up to go outside, but then give up. He would look up at us expecting a reprimand. We only assured him that it was okay and cleaned him up and wiped the floor. Karhu liked to lay out in the yard and the flies were drawn to his wound like it was a buffet table. Before leaving for work in the morning, I would put antibiotic cream on his wound and cover it with a towel. Otherwise, the flies would be all over it.

We had been giving Karhu 1-1/2 aspirin crushed up in a bowl of milk every twelve hours for the pain. It seemed to help. By Sunday night, the slight swelling I had noticed Saturday now involved his entire leg all the way to his toes and was feverish to the touch and purple in color. Very early Monday morning, Karhu woke us up crying in pain. He was on the floor by my husband's side of the bed. Fred stroked his side. The cries became whimpers but if the petting stopped, the crying would start again. After awhile I got up, grabbed and blanket and lay down on the floor beside him. For the next two hours I slept with him, my arm on his back. He was trembling with pain. The physical contact calmed him down and he went back to sleep.

In the morning, his leg was still the same. Fearing an infection had set in, I called the animal hospital. I told them my husband would be bringing him in that afternoon. After seeing Karhu, Dr. Rouse felt it was just the wound healing and draining but prescribed antibiotics to be safe. He was concerned about Karhu's appetite.

"He's eating like a horse," my husband answered. Unknown to us, our puppy had eaten very little during his three day stay and had lost five pounds. I guess hospital food isn't that great.

About the tenth day after the operation, Karhu's eyes were bright again and fully open. I swear I saw him smile at me, as if to say, "I'm going to be alright, Mom." He was getting around pretty good on the leg. The stitches had been scheduled to come out after ten days, but because of the Labor Day weekend and appointment problems, it was another week before Dr. Rouse could see Karhu. In the meantime, Karhu, with my help, had already taken out his stitches. After checking him over, the doctor was pleased with his progress.

At three weeks, Karhu was doing well. The pain was almost gone. He only yelped when he forgot and laid on his bad side.

Karhu had progressed so well at four weeks that Fred and I felt that a camping trip was in order. After the long three hour trip, Sultan and Heidi stretched their legs while we set up camp. Karhu was kept on a leash but he didn't seem to mind. It was just good to be out again. We let him play a little, then it was back on the leash. He was still napping quite a bit anyway.

At six weeks, Karhu went in for his final checkup and x-rays. The bones in his right hip were almost healed but would need a few more weeks to mend completely. Then Dr. Rouse gave us a scare. He had checked the x-rays for any signs of dysplasia in the left hip. If there were, he would have to perform surgery right away. The time frame for surgery was closing rapidly. But the left hip appeared normal and would probably stay that way.

Three months after the operation, Karhu's 3 inch long hair was growing back nicely. He and Sultan were body slamming each other into the ground, playing tag and "bring down the caribou." Heidi was usually the caribou. Our young wolf was running and jumping like a normal pup should. He was feeling great.

Now, four months later, Karhu is enjoying our weekly back country ski trips. Even after a long day of fighting deep snow, he has no pain or stiffness. Before the operation, a simple walk around the block left him limping in pain.

My words of advice to owners of Wolf Hybrids and dogs alike: be aware of the symptoms of hip dysplasia:

1. Unusual slowness in getting up,
2. Arching the back upon arising,
3. Reluctance to do the long forward and backward stretches a dog usually does,
4. Stiffness after exercise that doesn't go away with a short nap,
5. Inability to climb up on things.

If you see these symptoms, act quickly. Time is the biggest factor.

The optimum age for this type of surgery is seven to eight months, while the bones are still growing. It's too late at one year of age.

We were lucky. It was our active lifestyle that caused Karhu's condition to show up early. Usually the signs don't manifest themselves until later. Seek a second or third opinion until you receive the right answer. If you really love your pets, you will not let money be a factor.

Hip dysplasia does not have to be the life sentence of pain it used to be. I have read several articles on the subject, mostly by veterinarians and have never seen one mention the operation performed on Karhu. Even a recent article by Paul F. Bookbinder, DVM in the December issue of *The Wolf Hybrid Times* states "...there is no ultimate treatment for hip dysplasia." He is obviously wrong. That is why I wrote this story.

If just one dog or Wolf Hybrid's life can be saved because their owner read my story, then it was worth it. Not a day goes by that I don't give thanks that Karhu is okay. He went through so much pain before and after the surgery. Now he's free of pain and happy. I can't help but smile when I watch him jump to catch falling snowflakes.

There is hope for the victims of hip dysplasia. Karhu is living proof.

TOGETHER WE CAN DO IT

Dorothy Harpster

What's a success story? Like so many things, it is relative.

I was deeply saddened when I read what Runster and Karhu had been through for severe coxofemoral (hip) dysplasia.

Femoral head ostectomy, also referred to as excision arthroplasty, is usually not recommended by orthopedic veterinary surgeons for animals weighing over 20 kg (44 lbs.). The vast majority of Wolf Hybrids will surely have surpassed this weight by six months of age. It then becomes necessary to do the far more invasive surgery that Karhu had, the triple pelvic osteotomy. However, even with this procedure, not all dogs become pain free.

There is a third option for canines over ten months of age (to assure no further longitudinal femoral growth will occur). A total hip replacement is effective for disabling hip dysplasia on animals over 17 kg. (32-1/2 lbs). Currently, there is not a prosthesis available that will fit smaller dogs. For a good many Hybrid owners, this surgery may prove to be cost prohibitive. Total hip replacements are being done at all veterinary schools and many large surgical referral practices. Obviously, costs will vary considerably.

In Boston, at the Angell Memorial Animal Hospital, this surgery could be as low as $2,000; more likely $2,500. The implant (prosthesis) alone is $700 to $800. Post surgical care usually involves antibiotic therapy for four to five weeks, as there is great risk of infection with any implanted foreign body. Hospital compartmental confinement, then strict leash control are mandatory for four weeks. Then there is an additional two months of limited activity so that soft tissues around the surgical site are given adequate time to heal. Now, pray for no complications.

Dislocations are the most common complication encountered, of course. The most disastrous one is infection. This could then necessitate the removal of the cement and prosthesis. Then, if a second procedure is performed on a hip, there is now an even greater risk of infection.

I think that you'll agree that none of the above-mentioned surgical procedures are the answer. What is the solution? All Wolf Hybrids *could* be born with normal to almost near normal hips! In order for this to happen, anyone breeding their Wolf Hybrids must have these animals radiographed.

Your local veterinarian will have to tranquilize or anaesthetize your animals (pure dogs too) for ventrodorsal coxofemoral radiographs. Your veterinarian will be able to tell you if your animal is severely hip dysplastic. If this is so, no matter how superb you feel your 90 or 94 percent, narrow-chested, long legged, pale-eyed, rounder eared and wonderfully outgoing, friendly Hybrid is, he or she should not be used for breeding. You should neuter or spay your Hybrid and continue loving and enjoying your pet.

If your veterinarian thinks the x-rays look good, then they must be sent to OFA (Orthopedic Foundation For Animals), University of Missouri, Columbia MO 65211 to be evaluated. If your animal is under 24 months, the x-ray will be read by one veterinarian and it will be graded. Most high percentage Hybrids can't be mated until this age anyway. When your animal is over 24 months, additional x-rays should be sent. They will be read by three different Board certified veterinary radiologists and evaluated. There is an additional charge for elbows. (Readers should call or write for current charges). If your animal has normal hips at 24 months of age and is deemed satisfactory for breeding, he or she will then receive a breeder registration certificate. Your animal will have an OFA registry number. Costs will vary from veterinarian to veterinarian for the x-rays taken and the tranquilization or anaesthesia given.

Unfortunately, all breeders won't readily accept the added cost and trouble to do this. So if you are reading this and you plan to buy a Wolf Hybrid pup, then insist—no, *demand*—that the sire and dam are both OFA certified! If we all start demanding OFA parents of our pups, then more and more breeders will be forced to comply.

We absolutely need to start right now to try to eliminate hip dysplasia in Wolf Hybrids. No puppy should be born to suffer as Runster, Karhu and Tundra have. No animal should have to go through such painful invasive surgery and such an excruciating post-surgical period when it could be prevented.

The "Developmental Defects of the Canine Musculoskeletal System" article that appeared in December, 1989 stated that most of the diseases had not been proven to be hereditary. It also stated that overnutrition early in life, along with overexercise to *predisposed* animals could be the cause. It is important that you do not misinterpret that statement. *Predisposed* is the key word. I have spoken with Dr. George G. Keller at OFA. They keep statistics there on all breeds *(Canis familiarus).* I asked, because I knew there were breeds with extremely low incidences of hip dysplasia.

Take one of the very largest giant breeds—the Irish Wolfhound—only 5% of that breed is found to be hip dysplastic, while some smaller Spaniel breeds, pugs and Shih Tzus have extremely high rates of the disease. Originally, I thought that German Shepherds were the breed with the highest incidence. Not So. Right now, the Boykin Spaniel (South Carolina state dog) is number one. Limited numbers of dogs and extensive inbreeding contribute to the problem for this breed.

Dr. Keller told me the one breed that is virtually hip dysplasia free is the Greyhound. If the Greyhound had lameness or painful hips, it wouldn't be able to win races. Greyhound owners have culled the abnormal ones, leaving a breed with almost perfect hips. So where does this overnutrition play a part?

Obesity doesn't cause hypertension, diabetes or gall bladder disease in humans. But, if you are *predisposed* to any of those diseases, then obesity exacerbates each and every one. Overnutrition in young puppies (large or giant breeds) does not cause hip dysplasia. Surely the Irish Wolfhound and Greyhound puppies have their overindulgent owners and they don't have deformed hips. But for any animal that is predisposed to hip dysplasia, overnutrition and obesity will increase the incidence. OFA says hip dysplasia is breed dependent.

The musculoskeletal article ends with: "It is prudent for the well-informed breeder to assume conditions are inherited and refrain from breeding affected animals. Monty Sloan says: "Breeders, don't let the image of the 'perfect wolf' dissuade you from checking for developmental defects."

It is absolutely necessary right now for all breeders to have their breeding animals radiographically examined. If they are not normal, they should not be bred. Hip dysplasia free animals should be OFA certified. This will prevent the trauma that so many of our Hybrids have already gone through. Remember, if you DEMAND that a pup is from OFA certified parents, then the breeders will have to conform.

We should not anticipate that hip dysplasia in Wolf Hybrids will vanish overnight. Idealistically, if not one single hip dysplastic animal was bred again, there would be only a minimal number of diseased new pups born.

So let's get in the race and give those Greyhounds a real good run!

GLACIER—THE WONDER DOG

Paula Gmerek

Prior to reading Carla Childs' article, I thought we were the only ones who had a Hybrid with dysplasia. I was amazed to see the articles which followed. Like those owners, I wish to tell our story, which was at times very difficult, both emotionally and financially.

Glacier was our first Hybrid, a 68% MacKenzie Valley Timberwolf. We purchased him in August 1988 when he was three months old. We knew a little about Hybrids because our neighbor had one and we were really impressed with the temperament and sociability of their Hybrid.

We were like new parents and devoted all of our free time to our new Hybrid. Life was great until Glacier was eight months old. We began to notice that he was not running as much as before. When Glacier was nine months old, we were taking a lot of long walks and began to notice that once home, Glacier would sleep for about an hour and when he got up, he would cry. For awhile, we rested him and he seemed to get better again. As we began to increase his activity again, the pain would resume. We did take him to the vet, who said it was probably a pulled muscle.

Two more weeks went by and Glacier's pain was getting worse. We took him to the vet again, who this time took x-rays. He told us that Glacier had severe hip dysplasia—both hips—the worst he had ever seen. The vet then referred us to Dr. Mark Engen, an orthopedic surgeon. He also took x-rays and the outcome was the same. The dysplasia was so bad that an osteotomy was out of the question. What Glacier would need, if he was ever to resume normal activity, was a total hip replacement. The cost was going to be $2,000.00!!

Brian and I had a long discussion about our options and to us, there weren't any. We had fallen head over heels in love with Glacier and did not want to see him become severely arthritic in a year. So, at ten months of age, Glacier had his right hip replaced. He recovered well and after about two months, was running and bouncing around with his new brother.

It was five more months before we could afford to have Glacier's left hip replaced. The surgery went well, but instead of a large prosthesis, the surgeon had to use a medium one because the muscle holding the hip joint was too contracted to insert a large ball joint. The medium one fit perfectly and with this type of surgery, the muscle would contract further to hold the ball joint in the socket.

Things went well for the first three weeks. Then Glacier, who had been bearing total weight on the left hip, began to hold up the left leg. We took him to the vet, who said the hip had dislocated. He popped it back in and put the leg in a sling. One week after the sling was applied, the hip dislocated again. Back to the vet we went.

The surgeon was stumped. He took several more x-rays and consulted with the veterinary surgeon largely responsible for developing this surgical procedure about what to do next. Three weeks after the last set of x-rays, the decision was to replace the hip again, this time using a custom-made prosthesis which had an adjustable neck. The new prosthesis meant that the surgeon had to cut Glacier's femur bone in half in order to remove the old prosthesis and cement, insert a new prosthesis and wire the femur back together.

Glacier came through the surgery with flying colors. The surgeon informed us that he had to replace the socket as well and both the socket and ball joint were made out of space age material called titanium. Unlike Glacier's other hip, it was not cemented in, therefore requiring bone grafts around the prosthesis. The doctor told us that the recovery was going to be three months because with this prosthesis, the bone had to grow around it, so it was going to be more unstable during the healing period.

We had to put Glacier in a cage and keep him sedated for 1-1/2 months. He is doing beautifully and is in the fourth month post surgery. He is still taken for walks on a lead due to muscle weakness and at the end of April, will be able to run free. He has gained weight and is eating better than he has in a year. Although it cost us $5,200.00 and our families think we are crazy, we are grateful to have an 85 pound puppy again.

The girls at the surgeon's office call him "Glacier, the wonder dog," and he truly is.

ANTIFREEZE POISONING—AS DEADLY AS ARSENIC

Charles Nelson

Accidentally spill some antifreeze when you were changing or adding more to your radiator? Absent-mindedly leave a half-empty container sitting around where a curious animal might spy the brightly colored plastic and run off with it? Such unintentional carelessness could well mean the death of your wolf or Wolf Hybrid.

Less than half a cup of the rather good tasting liquid will kill an adult animal. According to Amy D. Shojai, 88% of the animals who have ingested antifreeze die, even those receiving immediate treatment. Worse, by the time an unsuspecting owner notices symptoms of antifreeze poisoning, it is usually too late to save the animal. If not humanely put to sleep, a natural death from such poisoning is traumatic and agonizing.

The initial symptoms of antifreeze poisoning are diverse and confusing. The animal may appear "drunk" or drugged, with an inability to coordinate voluntary muscle movement; vomiting, diarrhea and sometimes, convulsions. Loss of appetite, a rapid heart beat and panting respiration may be observed as early signs. The dog's abdomen may be painful as a result of renal swelling and fluid retention. Water intake and urine output will probably increase dramatically, due to kidney and liver damage.

The animal may seem to be recovering for a time, but after 24 hours, more serious and devastating symptoms appear and become increasingly more severe for as long as seven days before the animal finally falls into a coma and dies.

The moral of the story is to never drain, change, use or dispose of antifreeze in an area where *any* animal (or child) can get near it. A small dog, puppy or a cat can ingest enough of the poisonous fluid to kill it by playing with a piece of wood or a stone which is coated with it. (Little more than a tablespoon will kill a ten pound dog).

But if you should be unfortunate enough to find that one of your animals has ingested antifreeze, what should you do? 1) Call your vet immediately and arrange for him to meet you at the clinic (people rarely change their antifreeze during regularly scheduled clinic hours). 2) Induce vomiting if at all possible, and force feed as much water as possible to try to flush the poisons out before they can be absorbed. 3) If you are lucky enough to have activated charcoal on hand, force feed it immediately. According to Dr. Lawrence of the Texoma Veterinary Clinic in Sherman, Texas, "early administration of activated charcoal has seemed to improve survival." 4) Get the animal to the clinic as soon as possible, where it will immediately be put on a continuing infusion of fluids along with anticonvulsive and other medications.

None of the above will guarantee the animal's survival, (Dr. Lawrence points out that animals who have received no treatment for more than 24 hours after ingestion rarely survive), but following the above suggestions will definitely improve its chances of being among the 12% that survive poisoning from antifreeze.

The best cure - prevention, of course. Make sure that NO poisonous or hazardous materials are accessible to your animals at any time. It only takes a few seconds to remove them. It also takes only a few seconds for the animals to ingest enough antifreeze to kill them.

CHRONIC DIARRHEA

Elena Porter

I was reading a letter in the most recent *Wolf Hybrid Times* written by someone that had a pair of pups with chronic diarrhea. I thought I would share our experience, because our puppy almost died as a result of a genetic disorder in which the symptom was chronic diarrhea.

We bought a six week old female registered at 82% from a backyard breeder. This pair had never had pups before. Everything was fine for about a week and then she got diarrhea. No one got worried yet. Our vet treated her for various infections and we fed her on a diet of boiled chicken, rice and chicken broth. It didn't work. He treated her for some less common ailments. It didn't work.

At this point it had been about four to six weeks and the most alarming thing, besides the diarrhea, was that she was not gaining any weight. We went to a trained pathologist and he even put her on steroids. It didn't work.

After six to seven weeks of diarrhea, our poor pup looked like a tall skeleton. She could not concentrate on any training, since she was starving to death, literally. No matter how much we fed her, she continued to get taller but gain no weight.

Finally the pathologist found that she had a disorder called exocrine pancreatic insufficiency (EPI). This is a defect of the pancreas in which enough digestive enzymes are not produced and the pup can't properly digest the food it is taking in. No wonder she was starving! Luckily, EPI is completely treatable by adding an extract of pig pancreas to some moistened dry food. She is now about ten months old, weighs about 90 pounds and has no symptoms of EPI.

I would suggest to anyone whose pup or adult (EPI can have a late onset) has chronic diarrhea, to have them checked for EPI. They use a blood test in which, I believe, they look at serum trypsin levels. I hope someone else can learn from our experience. Consult your vet.

Editor's Note: *This is a rare cause, but should be considered if other treatments fail.*

CHAPTER VIII VACATIONS AND HOLIDAYS—HOW DO YOU MANAGE THEM?

Vacations and holidays can present a special problem for owners of Wolf Hybrids. How can you leave for a week or two and be confident that your animals are well-cared for? Can you take them along? How do you keep them safe, even if you are there?

This is an area of responsibility that is rarely thought of until the last minute—the minute you decide you absolutely *must* go somewhere else for awhile, or if you are suddenly called off on a business trip, or until holiday decorations are being unpacked. It is something which must be given consideration in advance in order to be sure that your animals receive the best care and nurturing while you are gone, or to plan for them to accompany you. It is something to be given consideration before holiday decorations come out of the boxes.

HAPPY VACATION

Bobbie Holaday

You can have a really fine vacation with or without your Wolf Hybrids, providing you make the proper preparations. I've found a few things that work for my Luke and Jato that I thought I'd pass along.

Some vacations are just for you and can't include pets, like when I take off for Club Med in the South Pacific. I'd worry myself sleepless if I left my precious boys under the supervision of some neighborhood kid. They might get loose and like most high content Hybrids, they are unpredictable about coming when called. Could I trust just anyone to check their stools every day to ensure their condition? Would they panic if Luke and Jato jumped on them? And the right thing to do? I have checked several animal boarding institutions and found that the Canine Country Club is one I can trust.

Canine Country Club is the oldest animal boarding institution in the Phoenix area. They have some 12' x 24' kennels where my inseparable Luke and Jato can stay together and still have room to romp, wrestle and rest. The kennel's staff are particular about the currency of shots and want to know the animal's history of ailments, likes and dislikes. They also happen to use the veterinarian that I do in event of an emergency.

The folks at the kennel weigh Luke and Jato when I take them in and give them a tick dip. Even though clients may dip their dogs at home, they do this to all animals as a precaution. They do check the animals' stools every day during their stay and check to make sure that they consume their meals.

No change in diet

They feed the dogs ANF kibble which I supplement with the raw meat that Luke and Jato need to get their full nutritional balance. I leave directions for giving my boys their daily vitamins. I also take large beef bones for occasional treats as well as eggs for every other day mixing with their food. The kennel staff will thaw out the frozen packages I've prepared and serve them at the times I've selected.

Living in Phoenix, the heat can decrease their appetites at times in summer and I've found a fairly sure-fire dish they can't resist. I mix cottage cheese with one or two eggs with their daily kibble ration. You can call this spoiling if you want, but they're worth it.

For a little extra charge, they will take Luke and Jato out together for a 20 minute run every day. They will also give them baths before I pick them up if I've so requested. About one hour and a half before I arrive, they administer the stress and travel sickness medications to ease the excitement when I appear to take them home. With all the attention Luke and Jato get from the kennel's staff, I think I'm the one who takes the separation the hardest.

Traveling together

On occasions where I can take Luke and Jato with me, and this means most of the time, I've learned to take precautions for stress and travel sickness. My Wolf Hybrids each have their own travel problems, but I've found medicine to ward off unpleasant results. I've found that any trip, whether out of town or to the kennels, causes fewer problems if no food has been consumed by my wolfies for several hours before we leave home.

Luke gets stressed from lengthy travel or change, so I give him a dose of Zema's Duo-Stress Capsules about an hour and a half before we drive to the kennels. For Jato, it's motion sickness and so he gets a dose of Shaw's Travel Sickness Capsules at the same time. These medications take effect best on empty stomachs. Our trips frequently mean an elevation change from 5,000 to 6,000 feet so they may appear tired for the first day.

Food for camping

Because our vacations are spent living out of my small RV and because my little travel refrigerator can't hold enough frozen meat for a week to ten days, I have to use mostly canned food plus their dry supplement. Being used to the good raw stuff, they turn up their noses at many canned foods. I found this out one recent vacation when I'd bought what I thought was the very finest quality of canned dog food only to have them push dirt into their dishes to bury them. So now I experiment at home ahead of time by trying out different high protein canned foods and I've found some they really wolf down.

I can stock up on things I know they will eat and not have to panic on my vacation because they're not eating. I do manage to fit in a few packages of frozen meat so they won't be entirely deprived of their favorite.

Affordable meat

Where and what do I purchase in the way of frozen meat? I've found that local meat packers will sell chopped up unsalable beef for 25 cents a pound. I can also get fresh kidneys or mature livers at 35 cents a pound.

They'll let me have beautiful, meaty beef bones at 15 cents a pound, less than what they are now at most supermarkets for the same quality. I can get chicken backs and necks at the local poultry packers for 25 cents a pound. By keeping my freezer packed with both beef and chicken, I can give them a variety.

I really love taking Luke and Jato off to some remote spot where they can frolic in some forest stream, splashing and swimming in the cool water. There they can hunt endlessly for rabbits and squirrels that they never catch. I can tell they love all the extra attention they get from me on these vacation trips, as we are together constantly. Nearby campers probably think I'm deranged when they hear me talking to dogs all the time, but of course, we all know how Wolf Hybrids understand us.

Be prepared

I did have a traumatic incident on one trip when Luke caught a toxic sting in the tongue and forced me to cut short our stay. His face swelled up and his eyeball was all white with no eye and he went into anphylactic shock. He was a pathetic sight most all the way home from eastern Arizona. I took him immediately to his veterinarian for treatment.

We never could determine what had stung Luke, but whatever it was, the effects almost eliminated his appetite and severely affected his activities for a week. The veterinarian has prescribed an injectable steroid and some follow-up tablets for me to purchase and have on hand in the event that either animal has a similar encounter in the future. Since Luke has such a violent reaction to a toxic sting, the next encounter could prove far more serious without immediate inoculation. Now, with the medication ready to administer, I can't prevent the sting, but I can cut down on the shock, stress and pain (mine as well as theirs). Fortunately, living on the desert, both Luke and Jato know what a rattlesnake is and keep their distance from those coiled warning signals.

This article is not meant as an endorsement or advertisement for the commercial companies or establishments I have mentioned. It merely informs the reader what I have found works for my Wolf Hybrids.

TRAVELIN'

It is that time of year again, when people are traveling, going to different rendezvous, seeing places, people and animals they've never seen before—or haven't seen for a long time—and their animals are going with them. Unless, of course, you're one of those fortunate people with a built-in "wolf sitter." Or maybe you're looking to ship an animal some place you're flying. Now, you've suddenly found out that none of the above are really such an easy task.

"Oh, my neighbors will take care of my animals." Will they? Will they make sure that the animals have food and that clean water is available to them 24 hours a day? Will they give them at least a few hugs or some attention, or will the animals destroy everything in sight trying to get out to find you or to escape "prisonerdom?"

On the other hand, if you take them with you, what are you going to encounter? Do you have to have permits in hand every time you cross a state line? What are you going to do with them at night, without them freaking out and destroying someone else's property? What are you going to do with them when you want to go "sightseeing or out for dinner?" If you are flying, and all of the above has already been taken into consideration, will federal regulations allow them to go?

By now a lot of the readers will have learned the hard way about traveling with animals—the way most of us learn. If you are traveling with a Wolf Hybrid in the car, you will, for example, have found that in addition to the spring shedding season, there is a lesser, but still heavy shedding season in the summer and that all those beautiful strands of wolf hair paste themselves to you beautifully when your skin is already moist with sweat. (The author has no suggestions for combatting this problem except for keeping your mouth closed).

As to overnight accommodations, many motels will allow you to have pets in your rooms, but any damage done by them will have to be paid for by you. Those foolish enough to leave animals locked in a strange room alone may return to find every drape, bedspread, pillow and towel in shreds on the floor and should consider themselves lucky if the window glass is still intact and the animal still contained inside.

This is not to say that every animal will react in this way, but like a frightened child, a frightened wolf or Wolf Hybrid panics easily and reacts. Far better, if the animal is to be left alone, to put him in a "wolf proof" portable kennel than to pay a motel several hundred dollars and perhaps risk losing the animal altogether.

Although there are no checks at state lines for animals crossing from one state to another, it is possible that if you were stopped for say, speeding or seat belts, that the officer would ask for current vaccination papers on the animals if he were suspicious that the animal might be part wolf and you were in a state prohibiting such animals. In order to go into Mexico or Canada, such papers, along with valid local tags and veterinarian's health certificates are required to cross the border.

It is advisable, no matter where you travel, to have all vaccinations current, along with heartworm prevention medication, for the safety of the animal itself. Different areas of the country have severe epidemics of corona, parvo, distemper, heartworm and lyme disease, to all of which you are unable to prevent your animal's exposure. Corona, distemper and parvo give no early warning signs. If you are on the road and miles from anywhere, the first signs you notice can mean death within several hours. Long term protection is essential.

If you are camping, be sure to check in advance on regulations in the parks or campgrounds where you plan to stay. In some states, Pennsylvania for example, you may not camp in the state parks with any pets whatsoever. You will have to settle for one of those private campgrounds that are often like miniature overcrowded mobile home parks, where your neighbors are less than 12' from your tent (and animal) and with no enforcement of restrictions on noise or intrusion on others' privacy or camping area.

In ANY camping area, you must keep your animals securely tied or contained in a manner that is both humane and that will not interfere with anyone else's enjoyment of the campground. This is only fair (and safer for your animal)—you wouldn't want anyone else's animals coming around and eating your food, fighting with your animal or frightening your children.

Flying with your animal? Maybe. Federal regulations prohibit the air transportation of animals in extreme temperatures, either hot or cold, because the cargo areas of the airplanes are not temperature controlled and not universally pressurized. The animals could not, in many cases, withstand such conditions.

Those are your only choices of animal transportation—by personal vehicle or by air. Except for special trains commissioned specifically for circuses, trains and buses can not and will not transport them.

But traveling with your Wolf Hybrid need not be an ordeal. As with any animal, just use good common sense. Carry fresh, cool water and make it available to them often. Feed them as you normally would after you have stopped for the night, and stop often for "pit stops." They need to stretch their legs and be able to walk around just like humans. Don't leave them alone while you go off somewhere else. And of course, don't pen ANY animal in a closed vehicle. Temperatures inside a vehicle, even in winter with the windows cracked open somewhat, can rapidly build to well over 100 degrees and their bodies, like yours, cannot withstand this. They will die of dehydration.

Using such common sense, there are many joys to traveling with your animals. They can show you things you would never have noticed before. The need for finding places to camp that are suitable for them allows you to experience sometimes remote and beautiful areas you would not otherwise see. By their uniqueness, they will introduce you to people you would not otherwise have met, who may become lasting friends. And they offer a degree of protection, in that most people will not knowingly approach a strange, large dog.

Given the choice and the ability, most Wolf Hybrid owners would take their animals with them when traveling and generally, once settling into the routine of it, the animals enjoy it immensely.

HOLIDAY TRAVEL

(OR SHIPPING ANIMALS LARGE OR SMALL)

Mel Tanner

My family wants to visit the grandparents over the Christmas school break but we can't leave our animals behind. We have no-one we really trust to take care of them. We want to take them with us and there are secure yards they could be in while we are visiting once we get there, but we could not fit them in the family car if we drove, and it would be a long drive. Is flying too costly and would there be any problems, or is there some form of ground transportation available?

You do not say where you are traveling to or from, any indication of the distance you would be traveling, nor the size or age of the animals. The best I can do is give some general transportation advice and hope that at least a portion of it is applicable.

First of all, let's assume that you are planning to travel a great distance and that air transportation would be the most logical means of transporting both your family and your animals to their destination.

Animals (of any species) being shipped by air, ride in the cargo section of the airplane. On some airplanes, the cargo section is pressurized—on others not. So far as I know, none are "climate controlled."

An animal is required to be kennelized in an air kennel approved by the particular airline you are using and unless you already have one, you must purchase it from the airline when you deliver the animal for shipment. By USDA requirements, the kennel in which the animal is shipped must be of adequate size to allow it to stand, lie down and turn around in comfortably - but not all terminals have a variety of kennels available and you should check with the airline before bringing the animal to the airport. (You must in any case make reservations for shipping the animal at least 24 hours before flight). You will not be able to ship the animal if you cannot provide a carrier meeting the airline's (or APHIS') specifications OR if they do not have one available for you to purchase. Within seven days of the shipment, you must obtain a veterinarian's certificate of good health and current vaccination status, and this certificate must accompany each animal.

Air shipment of animals by air is hazardous in both tepid and frigid temperatures. An animal switching from one climate to the other (particularly if young) may suffer drastically from the climatic changes, coupled with stress from the sudden changes in altitude, the generally stressful experience of being caged, the cages being moved by strange people, the loud drone from the engines and the fear engendered by all of the above. Additionally, if the crate is not the right size for the animal, it may be bumped and banged against the sides sufficiently to cause physical damage.

While puppies and smaller canines may not do much damage to an "airline crate," I have had adult Wolf Hybrids shipped to me via air arrive with only two or three bolts still holding the crates together and the area around the "front gate" was all but chewed through. I've also had airline personnel wanting me to "walk" strange and totally terrified animals down an escalator instead of helping to carry the airline carriers down the stairs or using the elevator.

If you are planning to fly, my suggestions are that you:

1) Try to get the animals on a direct, non-stop flight. This will minimize the stress of being moved in their carriers any more than necessary and also reduce the possibility that they might, like luggage, end up on the wrong flight. And obviously, it will reduce the amount of time in the kennel.
2) Try to find an airline that does provide pressurized cabins for live cargo.
3) If possible, leave early enough before the holidays that you can avoid the "holiday rush" and if possible, ride on the same flight as your animal.
4) Most airlines will not ship live animals when it is extremely cold or extremely hot at either the shipping point or the destination. Governmental regulations do prohibit this. Choose those that will not do so. Animals can freeze to death on airfields, and can freeze or die of suffocation or heat prostration during the flight in the cargo holds if the plane is delayed on the runway.

As stated above, the flight kennel must be sturdy, large enough to allow standing, turning around and lying down, and have the required ventilation. If in doubt, check with the cargo supervisor (not the ticket clerks). Best, if you can buy the kennel ahead of time and get the animal used to being in it, and to being carried around in the kennel before shipping.

If you are shipping puppies, two or three may be shipped in a crate which will still allow the above requirements and it will cost much less if you can do this. It will also stress the animals less as there will be companionship and the warmth of multiple bodies to comfort them.

Use a large marker pen and print your name, address and phone number, as well as someone the airline can contact at the animal's destination directly on the kennel in case the identifying documents are lost. Even if you are flying on the same plane, this should be done to make sure that there are no problems at the destination.

Although the airlines are regulated by APHIS, it is very hard for employers such as the airline industry to know about or control the behavior of some of their employees, although they are still legally responsible for them. One must realize that wolves/Wolf Hybrids may be a "curiosity" and that aside from the physical hazards mentioned above, wolves and Wolf Hybrids may be subject to special attention (which might be good or bad). If you have any complaint of alleged mistreatment or neglect of animals of any kind by airline personnel, report it immediately and follow through to the U.S. Department of Agriculture, APHIS, Animal Care Staff, 6505 Belcrest Road, Hyattsville MD 20782 (301) 436-7830.

Ground transportation

Choices of commercial ground transportation are very limited. Depending on who you are able to contract with and how far you have to go, it will also be very expensive. Generally, your only options are 1) self-transportation via rented vehicles which you pull yourself and 2) livestock or professional animal transporters.

The second is obviously going to be the most expensive, depending on the number of animals you are transporting, their size, and the number of miles you are going to transport them. If you use this means of transportation, you will generally need to contact someone dealing in livestock transportation or someone specialized in exotic animal transportation.

If you are moving them by commercial ground transportation for a great distance, you will need to make sure that water, food, and protection from the elements is provided. You should require that the driver call every twelve hours (at a minimum) to report on the care and condition of the animals, and the anticipated time of arrival.

Self-transportation

The very best way of transporting your animals, of course, is to take them yourself. With the increased emphasis on "do it yourself" endeavors, most of the equipment rental companies in the country now provide stock or animal trailers that you can rent and haul with your own vehicle for even a one-way trip. Some even have large, air conditioned vans and self-contained recreational vehicles.

With a small stock trailer, the animals can be in a larger area; huddle up if they want to; move back and forth if they need to; get all the air they need and, depending on the trailer you choose, all the protection from the elements that they need. There are few cars which can not utilize the temporary hitch provided to pull the trailer. If you are considering this option, you also have the option of "dressing out" the trailer, to make it fit the animals' needs. And you will have the animals with you.

Preparation

No matter which form of transportation is chosen, the animals should not be fed solid food for at least six hours before shipment, but water and exercise are essential, especially if it is to be a long journey.

If you are driving, take the animals out on the leash at regular intervals for exercise and elimination. Like humans, they need this too.

If you go into a restaurant to eat, make sure that they have proper ventilation (should you have opted for a closed vehicle). Even on the coldest winter days, the interior temperatures of a closed vehicle sitting in the sun can, within minutes, reach temperatures of over 100 degrees.

I would definitely suggest that if the trip involves driving less than eight hours, that you consider renting an open but fenced trailer and taking the animals with you. There will be much less stress on both you and the animals and it will be less expensive.

On the other hand, if there is any way that you can find a "dog sitter," it would be much better yet. The animals will have much less stress and even if they don't know their temporary caretaker, they will be in their own territory with the biggest stress being that of not "knowing" the person who is caring for them. Remember that you are not only talking about shipment one way. You are going to repeat the stress when you return.

SUMMER VACATION FROM A HYBRID'S POINT OF VIEW

While many Hybrid owners prefer to spend their summer vacations at home with their Hybrids, others choose to take them with them on trips, or to board them at a kennel while they travel. While obviously, the first choice is preferable from the Hybrid's point of view, all three choices have their own vacation-related problems which owners need to be reminded of.

The hazards of summer

With the onset of warm weather comes the annual onslaught of bugs and parasites. The most obvious insect of course, the flea, brings misery by the pound to the host. Fleas in and of themselves are bad enough—sometimes causing the poor host animal to scratch until raw and bleeding spots appear and the fur is pulled out. Often forgotten, but sometimes worse than the obvious and immediate effects of flea infestation and bites are diseases and parasites of which fleas are but carriers.

In the southwest, they are carriers of bubonic plague, a disease thought by people in most other areas of the country to be extinct. Perhaps even less known is that animals themselves can be afflicted by plague. The June 4, 1986, the AP news wire service reported a cat being treated near Santa Fe, New Mexico for bubonic plague. Fleas are also carriers of tapeworm and therefor are involved in worm problems.

Obviously, prevention is far better than after-the-fact treatment, but the poor owner is faced with a dilemma—which of the hundreds of products available should be used? Which is safe? Since most flea powders, sprays and dips contain Carbaryl, a toxic chemical which can cause tumors, stillbirths and birth defects, and flea collars containing tetrachlorvinphos are most often pulled off and chewed, especially if there is more than one animal, owners try many different methods. There is the ultrasonic flea collar, a flea collar which may or may not work, depending if you can keep it on them and undamaged. Naturalists claim that garlic and brewers yeast in the diet will develop a body odor which repels the nasty little fleas, as well as other insects. Others spray or dust in the immediate area where the animals will be, and still others prefer the pills available from veterinarians which have recently been approved for flea control.

There is no one succinct answer to what type of prevention is best. Were we all to live in the northern and higher elevations, the problem might be minimized, as fleas do not fare as well in cooler climates.

Other nasty insects more common to warm climates and areas with dense foliage are ticks and ear mites. While dips, powders and sprays are generally used as a prevention for them, a daily "once over" is also needed, with careful removal of any ticks found.

What about the ears?

When ticks become imbedded far down in the ear or ear mites are discovered, a daily, gentle infusion of the ear for several days with a common ear mite medicine such as Mitox or Oti-Care should loosen and kill them. Then, as the dog shakes its head, the movement will bring them out to the surface of the ear where they can be wiped away with a clean rag.

Never try to clean or scrape insects out with a Q-tip or other device. You may severely injure the animal. If the infusion does not work see your vet for treatment as soon as possible so that permanent injury to the inner ear may be avoided and the animal's discomfort ended.

Another consideration with ticks is that they can be carriers of Rocky Mountain Spotted Fever, a disease which is not confined to the west, and many other infections can occur, both to animals and people.

In the Atlantic coastal states, a severe and compounding illness to humans and animals alike is Lyme Disease. This is carried by "deer ticks" and the symptoms are both chronic and accelerating as time passes. Acute infections resulting from incomplete removal of the tick or ulceration of the bite spot can also cause other severe problems in both human and animal.

While summer is also the signal for worm infestations in animals, this problem has been covered in the book, *The Wolf Hybrid*. Suffice to say that people traveling to environments where heartworm, whipworm, etc., which may not be common to their own area, should consider beginning a prevention program with their veterinarian at least a month before traveling with their animals to areas where these parasites are common. It could save the life of your animal.

Black flies (or bot flies) and gnats create equal torture for Hybrids in many areas of the country as they settle on the tender edges of the outer ear, laying eggs and eventually causing a raw but crusty surface which can cover the whole outside of the ear. Again, there are many sprays which can be applied topically, but if you have more than one animal, you may again be looking at the problem of toxic chemicals because one animal will lick and clean the infected ears of the other. Plain old-fashioned Gentian Violet, an antiseptic, seems to be as effective and much less hazardous—provided you don't mind a multi-colored dog!

Through the south and west, all sorts of nasty little plants abound which can cripple your animal. (See "The Perils of Summer). Other plants, which when stepped on are very painful are the obvious cactus, goat heads (an insidious long-thorned, creeping ground vine which exudes a poison from the hard, pointed prongs of the seed pods, causing pain long after the barbs are extracted).

Turtles and frogs, if their skin is punctured, will cause illness and frothing at the mouth.

Grass awns—what are they?

Grass awns are the seed parts of a wild grass found everywhere in the United States, sometimes called "fox tails." As the grass plant matures, it develops a cluster of feathery fronds (See Figure 1). As the plant dries and turns a pale tan in color, the fronds toughen and the individual seed pods to which they are attached loosen and become a potential hazard for any animal that comes in contact with them.

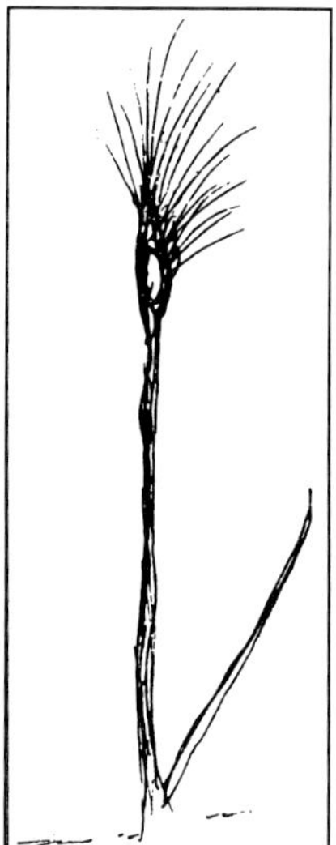
Figure 1

The plants are not poisonous. The danger comes when the needle-pointed awns become embedded between the animals' toes, in the skin, in the ears, eyes, or worse, in the throat or nasal passages.

As the animal moves around or tries to get the painful awn out on its own, the awns slowly work their way deeper and deeper into the tissue, sometimes moving as much as three inches from the original point of entry. Along with the pain comes the very strong likelihood of infection. Awns which are deeply embedded or which are located in the ears or throat must often be removed surgically.

Make a daily ritual of checking your animals' feet, ears and body, especially between the toes and in the ears. If the awn reaches the ear drum, the dog may become deaf.

Any time the animal licks or scratches a particular area continuously, it is a sign that there is something painful there that is bothering it and you should search diligently for the problem.

If all these hazards do not exist at home, you may be scratching your head about taking your animal into strange environments. If you are traveling, it would not hurt to do a little investigating about the areas you will be traveling through in advance—what the common clinical problems found there are, the types of hazards you may encounter—to make sure the animals have all the inoculations and preventive treatment available to protect them from the diseases and parasites common to the area

If you are to be near a lake or stream, be mindful that there may be tons of lost or hidden fishing hooks laying about. By streams and in certain mountain areas, there may be hidden, unsprung traps. There may be animals which can do terrible damage to the unrestrained or uninitiated "green-horn" Hybrid, such as raccoons, badgers, porcupines...and heaven forbid it should find a skunk!

If your animal does come in contact with a porcupine, either take it to a vet immediately or be sure to clip the end off each quill before you remove it, to release the vacuum in it. Then, after the quills are all removed, be sure to give the animal antibiotics for five days for the infection that is bound to follow. Best to take it to a vet for removal in the first place, however, as the pain of removal can be controlled with anaesthesia.

Protection of canine travelers

Even though some parts of the country are beginning to cool a little, owners need to be reminded that "cracking the windows" of the parked vehicle is simply not enough. Even in the dead of winter, the temperature inside a closed vehicle sitting in the sun can quickly build to over 100 degrees and the animal inside will dehydrate rapidly.

Far better, if you must leave it, to chain the animal securely in the shade with a bucket of water. Animals must be provided with fresh water often, especially if traveling, and not once a day as some seem to think is sufficient.

Another hazard which seems to be so common and somewhat of a "macho" practice is to expect animals to ride alone, unprotected, in the back of a pickup truck and to stay there, regardless of how the vehicle is being driven or what the temptations along the road may be. Too many animals are seriously injured and killed by their inability to keep their balance on sudden curves or stops, or when hanging over the edge of the truck bed barking at some attraction along the road; or just simply deciding to hop out while the vehicle is still moving. Even worse is the example of their being simply tied in the back of the truck and then either jumping or being thrown out of the vehicle and drug along by their necks.

If they are to ride in the back of a truck, they should be caged or else contained inside a camper shell with adequate ventilation. The latter, with opened but screened windows, would be preferable in any case because of the shade it affords.

Boarding

Depending on the length of the absence, many people prefer simply to board their Hybrids at their vets or a local kennel, feeling guilty all the while, though somewhat secure with the thought that they are safe and being cared for. This option may be alright for a day or two, but it carries negative sides too. The first must be fairly obvious: A veterinary clinic is constantly bombarded by germs and viruses from sick animals which are brought in to be treated, thus making it a veritable hotbed for disease, though all precautions are taken against it.

You must also realize that most kennels and clinics do not have the facilities or perhaps the time to exercise each animal individually on a daily basis. It would be a safe bet that except for the larger and more expensive facilities, the animals will not have the opportunity to run or move around much while you are gone.

The third drawback to boarding is the amount of stress on the animal from the sudden separation from the owner, the caging of the animal in a strange place and the probable change of diet. Many animals will not eat under these conditions and will become depressed and fearful. They develop diarrhea from the stress and the change, and often owners tend to blame kennel operators, claiming they have poorly managed and maintained their facilities, when actually, every precaution has been taken to prevent such physical problems clinically.

The other option

If the animal is to be left behind while the owner goes on vacation, even a short one, it is far more desirable to begin training a "babysitter" several weeks in advance of the planned absence. By getting the animal accustomed to the babysitter and that person taking over the feeding and exercise routines normally performed by the owner, it will gradually become accustomed to and comfortable with the babysitter, and the stress of separation from the owner should be greatly reduced. The stress will be reduced both by being familiar and comfortable with the babysitter and also by being in its own familiar territory and continuing in its own familiar routines.

Granted, the obtaining of a babysitter (perhaps combined with the function of housesitter) may be difficult, but is by far the better alternative where the Hybrid is concerned. The animal will adjust to a new person on its own territory much better than to a new environment in which it is confined, to which it is unaccustomed and in which there are so many risks. Theoretically, the owner will already have taken precautions against the environmental and physical hazards in his own facilities, and with proper maintenance by the babysitter, should have few concerns about the animal's welfare while he is gone.

There is no one answer—each owner must do what he or she feels is best, and while some animals travel very well with no problems whatever, others may not. The answer lies in the individual animal itself and the vigilance and protective measures taken by the owner.

One last thought is that whether traveling or at home, Hybrids should ALWAYS be restrained, whether by fencing in their own territories, chaining or kennelizing while traveling, or by keeping them strictly indoors or on a lead when outdoors. You do not want your animal to be run over by a car or truck; you do not want it to be shot or poisoned for running loose and destroying property or simply scaring someone camped next door; you don't want them falling or jumping out of moving vehicles; or simply to lose them because they have gone on a "free run" but not come back, for whatever reasons. As in any dealings with your animal, use your common sense. They depend on you for it.

HAVE SAFE, HAPPY HOLIDAYS

"Christmas comes but once a year,
and when it comes it brings good cheer."

Unfortunately, it also brings along other sometimes hidden, sometimes forgotten dangers. The season begins with the ritualistic getting down and unpacking the boxes of glittering, brightly colored decorations for the house. While Mom may be busily testing the tree lights and making sure there are enough hooks for the balls and other decorations, Dad and the kids are out tracking down that perfect Christmas tree. In the meantime, our favorite Hybrid is taking it all in—checking out the boxes and their shiny contents and stealing anything he can grab quickly enough to get away with.

But oh, the horrors what he has stolen can cause as it tries to move through his digestive system. Those beautiful shiny balls and light bulbs crunch up into slivers of razor sharp glass and metal and puncture the walls of the stomach and bowels. The chains of tinsel and strings of icicles wad up and become impassable balls gumming up or stuck in the intestines. The innocent-looking hooks that hold the ornaments on the tree may become imbedded in the mouth, throat, stomach or further down. No, Christmas decorations are not good things for dogs of any kind.

Finally, the newly cut Christmas tree arrives and is set up with lots of water to keep it fresh. That is important. Should the tree become dry, any little spark can set it off and turn the house into an inferno. One must be very careful about the placement of the wires to the Christmas tree lights. That little spark might very well come from wires that were chewed by puppies or adults, or are in the water. Wires, by the way, are also very convenient "strings" with which to pull the tree down.

It's also not a bad idea to cover and hide the water bucket. The tree is liable to come crashing down anyway if Fido decides he wants a drink out of it, again giving rise to the danger of a short in the wires, which may cause fire.

And oh, the lovely Christmas flower, the poinsettia. Basically a tropical plant of the spurge family *(Euphorbiaceae),* it is highly toxic if eaten and the plants should be kept well out of the range of the inquisitive Hybrid.

On a less gruesome note, but equally frustrating, the long awaited eve has finally arrived. The family is sound asleep and Santa's booty is piled high under and around the tree. This is *not* the night to be soft-hearted and let Fido sleep on the couch! The spot containing the festively wrapped presents which we spent so much time shopping for and wrapping, will very likely be a disaster area in the morning if you do, unless you do as one innovative subscriber did and fence the whole shebang in, tree and all!

A final reminder, for your own sanity—remember that all those fattening, oh so tasty goodies we all seem to gorge ourselves with during the holiday season are as tempting to canines as they are to humans. While cookies and other sugary treats taste as good to him as they do to you, sugar is definitely not good for canines. He may not understand why he can't steal bites the same as you do, but he sure won't understand why his stomach is upset or why he is suddenly put on a diet when it's all over!

"Kody and Tara." Photo: Cliff & Jody Hoydic

CHAPTER IX.
OWNERS SHARE THEIR THOUGHTS

In 1991, subscribers to *The Wolf Hybrid Times* received questionnaires requesting a variety of information ranging from why they acquired their first Hybrid, to how much money has been expended on their animals, to the owners sentiments on reintroduction and their personal involvements in efforts to protect or reintroduce the wolves in the wild.

Some of the respondents were resentful of the detail of the questionnaire, particularly where it requested information as to the amount of money spent annually for the care of their animals; however, most were very cooperative in at least attempting to estimate their costs for the pleasure of sharing a life with a wolf or Wolf Hybrid. The purpose for seeking such information was not to pry, but to be able to show those claiming lack of responsibility, that the majority of Wolf Hybrid owners are indeed very responsible and go to sometimes outlandish lengths to protect, care for and contain their animals and to show that Wolf Hybrid owners do care about and do support reintroduction programs throughout the continent. The following is a summary of the 481 responses received.

Why did you buy your first wolf hybrid?

The responses to this question were quite varied. 22% stated that they had acquired their first Hybrid because of fascination with the pure wolf, giving answers such as:

"Love, sympathy and empathy for wolves; need to perpetuate the symbol and value of the wolf in our country; encourage support of reintroduction."

"I love wolves but do not believe in keeping them as pets; the Wolf Hybrid is as close as I could get."

"Because of my love for the wolf and its nature."

"I have always been enamored with wolves; this was my opportunity to have a 'legal wolf."

Because they were lovely and unique; part of a lost and endangered heritage; because they were extremely sociable, more a companion than a pet; intelligent."

Nine percent reported acquiring their first Wolf Hybrid "accidentally," from an ad in the paper, or by adoption from an animal control or humane society facility. From persons such as those, we received responses such as:

"Acquired from a roadside zoo 15 years ago; from there, it never stopped."

"First one was given to me three years ago as a pup."

"Wanted something different to take the place of my Shepherd when he died. We sure got it."

"I do rescue work and two were turned in to me. I fell in love."

"Curiosity. I was looking for a canine companion who was more intense than a regular dog."

"Saw an ad in the paper; sort of an accident —a blessing in disguise."

Three respondents indicated that they had acquired their first Wolf Hybrid as a companion animal or mate for their domestic dog. Two of the three had the Wolf Hybrids spayed, feeling that it was not fair to the wolf. Only two respondents indicated that they had acquired their first Hybrid for "macho" reasons, such as:

"Don't know. Maybe because of the fantasy of owning a wolf or Wolf Hybrid, because everyone that I would meet would say that I was different and that I needed something different in a dog that would relate to my personality."

"Because I was attacked by an armed man at my apartment building. Saw an ad in the paper for a Wolf Hybrid. Wolf sounded pretty big to me and I'm so thankful I got my 'wolf'—he's my best friend."

62% of the responses indicated that they had given a great deal of thought and done a great deal of reading about wolves before purchasing or acquiring their Wolf Hybrid, with responses such as:

"We had been reading for years about wolves and Wolf Hybrids and decided to purchase one."

"Because of interactions with a neighbor's Hybrid and reading The Wolf Hybrid Times."

"Did a lot of reading and liked the attitude toward family, strength of purpose that appeared somewhat intellectual."

Some respondents had rather definite reasons for acquiring their first Wolf Hybrid:

"In 1945, another dog racer and I tried them in dog teams in Boulder, Colorado. A son brought the latest from Twin Cities."

"We were always interested in German Shepherds, but wanted to get away from purebred animals because of hip dysplasia and other congenital defects."

A very small percentage of the responses indicated that they had acquired their first because of falling in love with the puppy; many indicated that they had fallen in love with the parents.

Many indicated that they had been friends with Wolf Hybrids or wolves when they were younger, or that their spouses had. Nearly 85% indicated that their continued involvement with Wolf Hybrids had to do with the intelligence and/or uniqueness of the animals:

"I felt they were very unique animals, plus such smart and beautiful ones."

"A fascination with their beauty and intelligence. I wanted to try obedience training in a high percentage Hybrid."

"I wanted to give a good life to as many as I could afford; share their wild beauty."

"Because they are an intriguing and intelligent animal that I wanted to know better."

"On my first encounter with a Wolf Hybrid, I found them to have better temperaments than most dogs I'd known and they were beautiful; my husband's family had a Wolf Hybrid when he was young and he remembered him as a very special animal."

Several responses really caught our eye:

"Compassion, understanding, intelligence and most of all, love —from animal to me."

Pain in the wolves' eyes of abuse, i.e., puppy millers, abuse and stupidity of humans; education, understanding and love and affection to make the best better."

"Rescue of animals that were to be destroyed. I love all animals. I am hooked on Wolf Hybrids and want to make an uncertain future better."

How much money have you spent from January 1st to June 30th on feed for your wolf hybrid?

The responses to the foregoing question were incorporated with those to the following question of how many animals do you have, to derive the following.

While many respondents were feeding low cost diets to their animals, there were nearly an equal number feeding high cost feed or supplementing their diets with scraps from the butcher shop, table scraps or with fresh road kills or game hunted specifically for the animals.

The closest that we could estimate on actual expenditures for feed per animal per month, was an average of $42.00 per month (1991 figures) per animal (plus whatever supplemental food or vitamins were used). Using the figures provided, the survey indicated that in a six month period, approximately $335,120.36 was provided to the pet food industry by Wolf Hybrid owners alone.

Again, we would emphasize that *The Wolf Hybrid Times* is received by only a small fraction of the Wolf Hybrid owners/breeders in America, so this figure represents but a small portion of what is actually realized by canine food manufacturers from Wolf Hybrid owners.

How much *extra* money do you spend for feed for your wolf hybrids (as opposed to what you would have spent for other dogs you might have)?

77% of the respondents indicated that they would have used the same amount of food for any dog of comparable size. Most of those had either companion animals who were fed the same amount of food as the Hybrids. Others raised dogs of comparable size separately from the Hybrids and based their answer on this. However, approximately 40% indicated food expenditures of two to three times what they would have spent for other breeds of dogs because they were buying special kinds of food for them.

How much extra money do you spend when you want to travel to accommodate your hybrids?

67% of the responses to this question were similar to the responses that follow:

"Don't travel; we even pay a vet to come to our place when needed; $20.00 per visit."

"Don't travel—have too many animals and can't leave home overnight."

"My Hybrids don't allow me to travel."

30% indicated that they hired a "wolf sitter" (in most cases a family member) to take care of them if they did have to be away. 17% indicated that they had purchased and had specially equipped vehicles to go camping with their animals. 30% indicated that they would board the animals if they were to be gone for an extended period of time, listing rates of from $5.00 to $27.00 per day per animal. From those who did purchase special "accommodations" for their animals, these are some of the responses:

"We all camp together in the back of the truck."

"Bought a new Ford Explorer to travel with the animals."

"$200 kennel; $300 round trip air fare when he flies."

"$12,000—new Jeep."

"Pay a special rate at a motel to have them stay there too."

On a yearly basis, what is the biggest category of expenditures made on behalf of your wolf hybrid(s)?

Food by far outranked all other categories of annual expenditures, with many indicating elaborate diets. Additions to or replacement and repair of fencing were the biggest expenses of 15% of the respondents, while 24% indicated vet bills as their second largest expense. Time was the largest expense listed by 13% of the respondents.

"Hunting licenses. One or two deer taken a season is sufficient for my animals, but hunting isn't cheap."

"Changed my lifestyle and became a slave to a Hybrid."

We received many responses, such as:

"Toss-up between food and replacing destroyed items such as garden hoses, lawn furniture, croquet sets, shoes, siding on house, livestock, sprinklers, window, etc."

"Filling their bathtubs, giving them free-running time spent with them when I could be doing other things, but they need me more; all the film I use taking their pictures."

Do you support the reintroduction programs of the wolf in the wild; if so, how? Have you personally become involved in any of the preservation or reintroduction programs or efforts to provide proven rabies vaccines for wolves and wolf hybrids, and If so, how?

These were two separate questions propounded in the questionnaire. 99% of the respondents answered affirmatively, with most donations going to Defenders of Wildlife, The Sierra Club, The International Wolf Center, Alaska Wildlife Foundation, World Wildlife Fund, Green Peace, and Canadian Defenders of Wildlife.

Approximately 37% responded that they have been actively involved in educational programs themselves; approximately 77% reported that they have been personally involved by either writing letters to their representatives and congressmen in support of reintroduction or spending time distributing information about the programs.

The only "negative" answer we received was:

"No, I don't support it. I have seen too much irresponsibility with the government ranks to believe they could do a safe, sane job of reintroduction. The wolf would become nothing but a scapegoat for all the ranchers surrounding the boundary who happen to 'misplace' livestock."

One stated:

"Yes, but the public must be educated that wolves are not dangerous to their health. The wolf was—and should be—an essential part of our ecosystem."

If the laws in your state were to suddenly change and forbid ownership of wolf hybrids, what would you do?

While 83% of the responses indicated that if all else failed, they would move, most were quite vocal in expressing their sentiments.

"Everything and anything to save my animals, but if it came to the point that moving was the only way to save them—we would move."

"I would NEVER give up my animals. I would hope to be grandfathered in. I have secure kennels that could be checked—or I'd move!"

"I have already moved once because of this very thing."

"Put a 'For Sale' sign in front of the house and move to a state where the laws were reasonable."

"Take my wolf to the state capitol steps with all the signatures from the

people in my home town that Simba has helped. Simba served as the 'Public Library Wolf' for the 'Wolf Pack Reading Program,' has visited senior citizen homes for petting and loving by the older residents and has made personal appearances for young people's groups. If that didn't work, I'd move to another state!"

32% said they would lie about what the animals really were:

"Good question. I'm in the Army, so I can't relocate at will. I suppose I would attempt to hide them or disguise them (like shaving them to look like poodles)."

"I would keep my animals. They are not registered as Wolf Hybrids. Their vet has them recorded as Malamute. The state would have the burden of proof, which I would fight with all my resources. I'm not above lying to protect them."

"I'd reflect for awhile and weigh if I could win in the courts. If not, I would probably own a 'Malamute mix."

"Raise hell and hide my dog! and probably lie about him being any kind of 'wolf,' because he doesn't look like a wolf—he looks like a Lab."

17% indicated they would fight such a prohibition in court before taking other steps."

"Contact the governor's staff, as well as legislators and animal rights lobbyists."

"Anything we could to keep our animals, including calling them Huskies. We can't give them up! The smart move is becoming involved before the laws change, which we are doing."

"Change them via the courts. I am an attorney."

"Take a stand! I own my land and my home. As far as I'm concerned, no-one's touching my dogs. They are like children to me."

"Fight in court; move to lots of acreage and post no trespassing signs."

"Keep my wolf and shoot anyone who tries to take her."

Only two respondents indicated they would give up their Hybrids and try to find them a good home elsewhere.

What is the most outrageous thing that you have done for your wolf hybrids and how much did it cost?

While 87% of the responses involved expenses for feed, fencing, veterinary expense or costs to purchase special transportation, there were so many so unique that they are singled out below.

"Purchased a new Chevrolet Silverado and camper top equipped with specially made extra-security locking windows so we can take trips and our wolf can travel in complete comfort, costing approximately $24,000. I also to plan to purchase a spinning wheel ($800) to spin the fur for a 'genuine wolf fur cape' - another generous gift of love from my forever friend."

"I made them a cave and a wolfy condo."

"Construction of facilities better than the local zoo has; also, moving out of state to a legal area."

"We fed them the wrong type of food. Cost: $200 in vet bills."

"Last year, a road went in by my property, bringing close neighbors. We dropped everything and built a two acre compound—100' of kennels, houses; layed ground wire, etc. For the year 1990—$7,000."

"Spent $6,000 on hip replacements bilaterally x 2 for my oldest."

"Dug and built by hand a wolf pen, complete with an underground wolf den. Cost: $2,000, plus probably the same amount in treating them as our children."

"Interrupted our wedding and honeymoon in Florida (we were in Massachussetts) in order to fly home the moment she jumped out the window to evade the dog sitter. She ran up into the woods and howled plaintively—she was anxious to be found. $3,000 and a year of heartache. We survived and love is stronger all around. Still not married yet, though!"

"I always had those blue vinyl kiddie pools for my canines. My first Wolf Hybrid (female) enjoyed it and did not chew it or destroy the cheap, inexpensive pool. When the male saw his second summer (age 15 months), he promptly pulled the just-filled pool on its side, emptied it and ran around the yard throwing it about, much like a frisbee. I knew right then and there that it would never work. What I needed was an in-ground pool for them.

I don't have to tell you the look my husband gave me when this was suggested. So, I ran around to garden centers and purchased a black fiberglass pool/pond used for goldfish, outdoor ponds and water fountains, etc. My son dug the hole, and my wolf plumber (he has Hybrids too) laid plastic pipe to outside of the fence under the ground. He put a bathtub type drain in it with the valve outside the fence. We cemented the pool in the ground and covered up the pipe. Since then, my husband purchased flagstone and made a cement bed around the pool to lay the stone in. Now, there is a nice little patio around it.

I clean it once a week and refill it. They don't destroy it and it is permanent - I hope! P.S. The shape of the pool is that of a pork chop!"

"I bought a 12' x 8' child's playhouse ($500) complete with front porch, windows with screens, front steps, etc. I put this house inside their pen so they have a place to get out of the rain. We also took our two wolves into an Olan Mills photo studio for family portraits. What a disaster!

Although not asked on the survey, I thought you might be amused to hear of the various things they have destroyed in the past four years:

Two oriental carpets
Lawn Mower
Barbecue grill
Wife's right front hubcap
Two doors

Six vents in foundation
Various articles of clothing
Case of good wine
Army hat
Two lamps
One large stuffed chair
The side of one dresser
Foot stool
About ten square feet of carpet
One window screen
3 wooden bowls
10 lawn sprinklers
Wiring to our heat pump
20 feet of insulation
45 feet of buried hose from sump pump
three sheets of dry wall
One wallet with money, cards, IDs, etc.
3 silk flower arrangements
$500 masonite siding (house exterior)
Eight large (10') rose bushes
One 40 foot poplar tree
Drivers seat in 1986 Isuzu Trooper
One fan
Two weight lifting benches
TV and phone cables (about ten times each)
One set of curtains
Untold number of neighborhood frisbees, soccer balls and baseballs
Hundreds of shoes, magazines, books and garden hoses.

APPENDIX A

The Wildlife Education and Research Foundation
(corporate office)
P.O. Box 1423
Gallup NM 87305
(505) 863-5408

ANNUAL REPORT
December 31, 1994

Statement of Purpose

The Wildlife Education and Research Foundation (WERF) is a non-profit corporation established in 1986. WERF strives to educate the general public about remaining wildlife, with particular emphasis on endangered and threatened species, such as the wolf. WERF works to encourage, conduct and provide opportunities for research relating to the endangered and threatened North American wildlife, and as possible, to provide facilities for rescue and rehabilitation.

Foundation Programs and Objectives

Rabies Research Program

The goal which has been given the highest priority is the program to test and prove the efficacy of existing canine rabies vaccines in wolves and Wolf Hybrids. The U.S. Department of Agriculture requires that complete testing be conducted by veterinary rabies researchers whenever a change in label claims are to be made, just as if the vaccine were first being developed. The *Code of Federal Regulations* details very specific procedures which must be followed and the vaccine companies are monitored very closely during this process.

The program is costly and time consuming, and WERF has committed to raising as much funding as possible to aid recalcitrant manufacturers in obtaining approval for such a label claim addition. To date, the Foundation has amassed $50,000 for this program—about 1/8 of the total amount required. Fundraising efforts continue, as well as continuing liaison with federal regulatory agencies, rabies research laboratories and manufacturers. The program took a giant leap forward with receipt of a large private grant specifically designated for this program.

Through negotiation with the proper USDA officials, the initially required number of animals has been reduced, which has enabled the Foundation to restructure its original cost estimates.

Revised architectural and site plans have been finalized for the holding facility and as additional funding is received, bids will be solicited for construction.

Educational Programs

The Wildlife Education and Research Foundation collects data annually from state regulatory agencies regarding private ownership of wolves and Wolf Hybrids. The state regulations are summarized directly from the statutes and the resulting compilation is distributed to state agencies, humane societies, animal shelters, owners and other interested individuals without charge. Information on local and municipal regulations is gathered and distributed as available.

The Foundation has prepared a substantial "Fact Sheet" regarding Wolf Hybrids. The document is approximately 40 pages in length and all statements made are heavily documented. It is hoped that this document will help to combat some of the false claims and misinformation being generated and distributed by some organizations and individuals seeking to eradicate this breed of canine.

The WERF staff continues its extensive liaison work with state and local regulatory agencies which are under pressure to develop regulations regarding wolf and Wolf Hybrid ownership. The Foundation has been successful in negotiating reasonable regulations and/or preventing adoption of prohibitive regulations in seven states through its educational programs. WERF feels that open liaison and communication with these agencies is essential to promote understanding and cooperation between two often opposing factions.

Rescue and Rehabilitation

At year end of 1993, a large private grant was received with two very specific designations as to how the funds were to be used. A large portion of the funding was designated by the donor for development of a rescue and rehabilitation facility for wolves and Wolf Hybrids. The Foundation acquired a facility known as Candy Kitchen Rescue Ranch near Ramah, New Mexico and procured groundwork, fencing and labor to construct large animal enclosures, shelters, equipment and provided veterinary care and maintenance for the animals. The refuge now houses some forty displaced animals.

As specified by the donor, the Foundation has managed the refuge fiscally, and will provide technical assistance necessary for the refuge to become autonomous and self-sufficient. It is anticipated that this goal will be accomplished by early 1995.

Seminars

A full-day, video-illustrated seminar has been developed for presentation to animal shelters, regulatory personnel and interested individuals. The seminar involves audience participation and focuses on the impossibility of accurate identification, the proper management of wolves/Wolf Hybrids, adequate containment, legalities of regulation and many other topics. While none of the full day seminars have been presented this year, portions of the material and video have been presented to legislative and regulatory agencies and to groups of private owners. These presentations have been very successful in destroying preconceived misconceptions of those attending.

Future

The Foundation will continue the programs already initiated. Major emphasis will continue to be placed upon the rabies research program, as we strive to resolve this question. As rabies epidemics rage up the East Coast and through southern Texas, this is a question that must be resolved as quickly as possible. The numbers of wolves and Wolf Hybrids in private ownership continues to increase annually and debate against wolf reintroduction always comes back to the issue of transmission of disease to the domestic population of canines and livestock and their owners.

"Cheyenne Kyle" Photo: Bill Kyle

BIBLIOGRAPHY

Brackett, Lloyd, "Planned Breeding," *Dog World,* Chicago.

Frankling, Eleanor, *Practical Dog Breeding and Genetics,* New York, Arco Publishing Company

Onstott, Kyle, *The New Art of Breeding Better Dogs,* New York, Howell Book House.

Prendergast, Dorothy, *The Wolf Hybrid,* Rudelhaus Ent., PO Box 1423, Gallup, New Mexico 87305, 1989.

Vainisi, et al, "Nutritional Cataracts in Timber Wolves," *Journal of the American Veterinary Medical Association,* Vol. 197, No. II.

The Wildlife Education and Research Foundation, P.O. Box 1423, Gallup, New Mexico 87305

The Wolf Hybrid Times, Rudelhaus Ent., P.O. Box 1423, Gallup, New Mexico 87305.